A HISTORY

OF THE

STATE OF DELAWARE,

FROM

Its First Settlement until the Present Time,

CONTAINING

A FULL ACCOUNT OF THE FIRST DUTCH AND SWEDISH SETTLEMENTS,

WITH

A DESCRIPTION OF ITS GEOGRAPHY AND GEOLOGY.

BY

FRANCIS VINCENT,

WILMINGTON, DEL.

PHILADELPHIA:
JOHN CAMPBELL, NO. 740 SANSOM STREET.
1870.

Henry B. Ashmead, Book and Job Printer,
1102 & 1104 Sansom Street, Philadelphia.

ADVERTISEMENT.

From the smallness of the State of Delaware, both in population and territory, and the few (even of Delawareans) who manifest any interest in its affairs, the author has been compelled to issue this work in numbers of thirty-two pages each, at thirty cents per number, supposing in that manner it would be placed more easily within the reach of a greater number of subscribers. They will be issued about every three weeks, and can be procured either of the author, at Wilmington, Delaware, of Mr. John Campbell, No. 740 Sansom Street, Philadelphia, or of the bookstores in the City of Wilmington.

By forwarding the money by mail to the undersigned at Wilmington, Delaware, the numbers will be sent in any direction free of postage. Back numbers will always be furnished. The usual discount made to dealers.

As many of the subscribers to the work appended their names about four years ago, expecting to be supplied with the History at that time, it may have escaped the memory of some of them. Should any wish to cancel their subscription from this cause, they will please, by some means, inform the author.

Francis Vincent.

Wilmington, Del.

ERRATA.

The following are the principal errors, which have escaped the observation of the author and printer.

In	page	14,	line	31,	for	seven	read	eight.
"	"	25,	"	23,	"	White	"	Vine.
"	"	27,	"	5,	"	twenty	"	fifty.
"	"	32,	"	10,	"	Mill Creek	"	Christiana.
"	"	168,	"	1,	"	Peter	"	Gillis.
"	"	219,	"	30,	"	Anundsen	"	Amundsen.
"	"	290,	"	4,	"	Governor	"	principal officers.
"	"	300,	"	27,	"	Emanuel	"	the Presbyterian.
"	"	319,	"	22,	"	Adrian	"	Cornelius.
"	"	392,	"	19,	"	1769	"	1679.
"	"	462,	"	28,	"	1868	"	1851.
"	"	463,	"	1,	"	1699	"	1679.
"	"	476,	"	21,	"	1860	"	1840.
"	"	"	"	23,	"	Eleanor C.	"	Eleanor.

In speaking of the descendants of Vincent Loockermans, no mention is made of the youngest daughter of Sarah Loockermans, who married the Honorable Nicholas G. Williamson, formerly Mayor and Post Master of Wilmington. Her name was Lavinia Rodney. She married William O. Bateman, who was, for twenty years, a prominent member of the Philadelphia bar, and the Democratic candidate against Judge Stroud in 1861.

PREFACE.

THE author has no other apology to make for undertaking the present work, than that in the course of two hundred and thirty-eight years, (the period of time that has elapsed since the first settlement was attempted of the territory that now constitutes this State,) no other person has thought proper to write it before him. Delaware has a history that every citizen should be proud of. It is many years older than that of her great sister, Pennsylvania. Yet how few Delawareans there are who can tell who settled or governed it, who fought its battles in the Revolution, who passed its laws, what they were, or the circumstances under which they were enacted, or the social and political changes that have from time to time occurred within their State. The want of a relation of these transactions in a book easy of reference, has made this ignorance of our State affairs a general ignorance. This work is designed to remedy this, and to place it in the power of all Delawareans to obtain a knowledge of the past occurrences of their State, that may desire it. All the author aims at, is to plainly, truthfully, and succinctly, detail what has transpired, or may in any way relate to the history of Delaware, in a manner that may be understood by all.

The difficulties of writing a correct History of Delaware can hardly be conceived by any who have not undertaken it. In addition to the extreme lack of historical interest in relation to their State manifested by many, even of our best citizens, no care has been taken of our records. In 1722 they were all destroyed by the burning of Major John French's house, at New Castle. In 1777 they were captured and carried to New York by the British after the battle of Brandywine. Many of them were never returned, and what were, were stowed away in an outhouse, and afterwards nearly all burnt by a gentleman's servants, (to light fires,) who were ignorant of their value. Others have been lost by the carelessness of our different state and county officers. In moving the Kent county records from the State House in Dover (where many of the officers of Kent county were) to the newly erected county buildings a few years since, a large quantity of valuable matter that would have thrown light on our State history was carted away, and cast out as rubbish. A number of valuable letters and manuscripts belonging to Thomas M. Rodney, Esq., consisting of letters of his great uncle, the celebrated Cæsar Rodney, in relation to the days of the Revolution, were stolen some few years since. Many of our former most distinguished families have now no representative left in the State, and their papers that would have thrown light on our local history are not to be found. The descendants of others who reside here have parted with every manuscript and letter, having in many cases presented them to New England autograph collectors, amongst whom such things are preserved and valued. No care has ever been taken in our public libraries to preserve any works in relation to this State, and every rare work, not purchasable, the author has been compelled to procure from the libraries of Philadelphia and New York. From this our readers can imagine the difficulties there will be in writing a history of our State; especially after its grant by the Duke of York to William Penn—a few years after which grant we first had an independent government—and our affairs ceased to be registered anywhere out of our own limits.

This State having been first discovered by the Dutch, and the first settlement made by them and the Swedes, its early history must necessarily be found in those languages. These two nations inhabited the Delaware between them, and struggled for mastery, until finally the former reduced the latter to subjection. The Dutch officials on the Delaware sent a regular account to New York of every matter of interest. The New York officials sent copies of those accounts to Holland. The Swedes on the Delaware sent accounts to Sweden. These several accounts, both Dutch and Swedish, have many of them been preserved either in the original manuscripts, or by record in books, and it is from these records and cotemporary works, written by Dutch and Swedish authors, that we are enabled to get a minute and circumstantial account of our early history.

The Dutch records by the liberality of the State of New York have been translated into English, by Broadhead and O'Callighan, and published in thirty large volumes. Among these records are the letters of Hudde and Beekman, both of whom minutely record the occurrences on the Delaware. The correspondence of the latter, who was governor of the territory of the northern side of the Christiana, is especially valuable, containing, as it does, a succinct record of the events in the territory that now comprises this State from 1659 to 1664. The Swedish documents, from which we extract our principal information in relation to Swedish affairs, were furnished by Mr. Russell, our former minister to Sweden, to the American Philosophical Society of Philadelphia. They were translated and published several years ago in the Register of Pennsylvania. One of the most useful (though also, in some instances, one of the most inaccurate,) of Swedish works is that entitled, "*A Short Description of the Province of New Sweden, by Thomas Campanius, of Holme*," printed at Stockholm in the year 1702, under the patronage of Charles the 12th.

Delaware, from the time of the subjugation of the Swedes, in 1655 to 1682, (when it was conveyed to Penn,) being but a *sub* government to New York, her history up to that time, as well under the English as under the Dutch, must be looked for mainly in the records of that State. Nicholas, Lovelace and Andross, the deputy governors under the Duke of York, (who was the proprietor of this State, and afterwards granted it to Penn,) had regular accounts sent on to them of the affairs on the Delaware, where they were carefully recorded. Our own records (with the exceptions of a few deeds and wills) do not extend further back than 1674, when Andross was governor. They are contained in three books in the Prothonotary's office at New Castle, and at least one of them is a copy from the records of New York.

The author designs that this work shall contain a narrative of Delaware events from its first discovery until the end of the year 1869. The plan of the work will be to give every public event, the essence of all important laws passed, the names of the governors, the legislators, and other important public officers. The different census, and the number of votes cast, and the majorities at the various elections, and the principles and objects of the various political parties that have existed in the State. With this declaration of the object of the work, he issues his first volume, which is now written in hopes it will meet the approbation and patronage of the citizens of Delaware. As the second volume is not yet written, he should be pleased if any who have any letters, manuscripts, documents or books, that will throw any light on the past history of Delaware, would loan them to him for a perusal. They will be faithfully returned.

Since the above was written the author has added two chapters to the work, more than he at first intended. The first giving an account of the boundaries of the State and its Counties, and their extent, as well as a short description of its surface, harbors, bays, rivers and creeks; the other, a description of its geological formation. He has done this, supposing that the information conveyed would cause the historical portions of the work, when they came to be read, to be better understood.

HISTORY OF THE STATE OF DELAWARE.

CHAPTER I.

The Radii from New Castle—Mason and Dixon's Line—Jurisdiction over the Delaware for Twenty-four Miles—Latitude and Longitude—Counties of the State—Its Hundreds—Estimated Extent of its Territory—Boundaries of Counties—Northern Part Hilly—The Ridge—Cypress Swamp—The Forests, White, Black, Spanish Oak, the Bark—Game—Mocking Birds—The Rivers, Harbors and Streams—Naaman's Creek, Brandywine, Christiana, White Clay and Red Clay, Mill and Bear, Red Lion, St. Georges, St. Augustine, Silver Run, Duck and Little Duck Creeks—Kent and Kelley's Islands—Dona River—Port Mahon, Little Jones', Mispillion, Broadkiln and Lewes Creek—Lewes Creek filled up—Cape Lewes—Rehoboth and Indian River Bays—Indian River Inlet—Lewes, Middle, Herring and Guinea Creeks—Long Neck—Burton's Island—Pepper, Vine and White Creeks—Kedging of the Shallops—Fresh Pond—Salt Pond—Manufacture of Salt—Assawaman Bay—Fenwick's Island—Assateague Island—Fish and Water Fowl of the State, the Crocus, the Sheephead, the Drum, the Man-nin-nose, the King Crab, the Curlew—The Nanticoke, Broad and Pokomoke Rivers—Rivers rising in the State flowing into the Chesapeake.

The boundaries of the State of Delaware are as follows: first, a circle drawn in a radius of twelve miles from the Court-house at the centre of the town of New Castle, commencing (we will say) at low water mark on the shore of New Jersey, north of New Castle, thence extending over the Delaware, and following its circumference until it again touches the shore of that State south of its radius of twelve miles from New Castle. This circular boundary on the north gives Delaware sole jurisdiction over the Delaware River and Bay from low water mark on the Jersey side, over a mile north of where Naaman's Creek, on the western or Delaware side, flows into the Delaware, extending southward twenty-four miles to a place a short distance north of where Silver Run Creek enters the Delaware from this State, or about a mile south of where Alloway's Creek flows into it from the Jersey side. Within these boundaries are comprised the islands of the Pea Patch, on which is erected Fort Delaware, and Reedy Island. Below this circle the jurisdiction of the State extends to the middle of the bay, as far as Cape Henlopen, where it flows into the Atlantic Ocean. It then extends along the Atlantic Ocean to a point at Fenwick's Island, in about 28° 20′ north latitude. The line of the State then extends westwardly thirty-four miles, three hundred and nine perches (being exactly half the distance between the ocean and the Chesapeake Bay). The State boundary then runs by a right line nearly due north at a tangent until it reaches the western part of the periphery of the circle, twelve miles from the Court-house at New Castle. It contains within its limits 2002·6 square miles. The State is situated in latitude from 38° 28′ to 39° 47′ north, and from longitude from 74° 56′ to 75° 46′ west from Greenwich. Its physical boundaries are as follows: on the north by Pennsylvania and the Delaware River, on the south by Maryland, on the east by the middle of the bay and river to twenty-four miles from the State's northern boundary, from thence by a line of low water mark on the Jersey shore to the radius of twelve miles north of New Castle; and on the west by Maryland, and by Pennsylvania to the periphery of the circle from New Castle, where she connects with the State of Maryland. This circular boundary of Delaware causes the entrance of Pennsylvania between Delaware and Maryland in the shape of a long narrow wedge. The length of the State is ninety-five miles. At its southern boundary it is nearly thirty-five miles in width, which width is hardly diminished for about twenty-six miles, or to Cape Henlopen. But from Cape Henlopen to its northern boundary, from the Delaware flowing in a southwest course, it diminishes in width until it reaches its narrowest part in the neighborhood of Red Lion Creek, in New Castle county, (where its breadth is not over ten miles,) when it again widens until it reaches the breadth of twelve miles from New Castle Court-house. The line that divides Delaware from Maryland is a part of the celebrated Mason and Dixon's line, run by Charles Mason and Jeremiah Dixon in 1762, (of which we shall speak more hereafter in its proper place,) to separate the territories of Lord Baltimore and Thomas and Richard Penn, sons of William Penn. This Mason and Dixon's line was popularly supposed to be the boundaries between the free and slave states. But this was a popular error. Slavery existed in Delaware, which is west of this line, until abolished by the fourteenth amendment to the constitution. The mistake occurred, we suppose, from the line when it passes the periphery of the circle from New Castle and reaches the boundary between Pennsylvania and Maryland, instead of running north and south, as between Maryland and Delaware, runs due west. So that Mason and Dixon's line was the boundary between slavery and freedom when it ran east and west, between Pennsylvania and Maryland, but not when it ran north and south, between Maryland and Delaware.

The State is divided into three counties, viz: New Castle, Kent and Sussex, and each of these counties are subdivided into hundreds. Hundreds are the old English subdivisions of counties, and Delaware is the only State, it is alleged, in which they exist in the United States. They were supposed to have originated with Alfred the Great, one of the old Anglo-Saxon kings who ruled in England A.D. 877, or 992 years ago. But they are now known to have been in existence before his time. They derive their name from having originally, when instituted in England, contained one hundred families. New Castle county contains ten hundreds, viz.: Brandywine, Christiana, Wilmington, (the city of Wilmington, which, by law, is a hundred in itself,) Mill Creek, White Clay Creek, Pencader, New Castle, Red Lion, St. Georges and Appoquinimink. It contains 424.02 square miles. New Castle is the county town.

Kent county contains seven hundreds, viz.: Duck Creek, Little Creek, Kenton, Dover, North Murderkill,

South Murderkill, Milford and Mispillion.[1] It contains 613·06 square miles. Dover is the county town, and capital of the State.

Sussex county contains eleven hundreds, viz.: Cedar Creek, Broadkiln, Georgetown, Nanticoke, North West Fork, Broad Creek, Little Creek, Dagsboro', Baltimore, Indian River, and Lewes and Rehoboth. It also contains 964·08 square miles. Georgetown is the county town.

There are various statements published of the size of Delaware, nearly all of which differ; but the account we have given is based on the last survey made by D. G. Beers, for the publication of the State Atlas of Pomeroy & Beers, in 1868. The survey of Rea & Price for the State Map in 1850 gives the area of the State at 2,221 square miles. But they vary in their own calculation, for they give the number of acres contained in the counties of the State 1,300,250; this, divided by 640, the number of acres in a square mile, would make the State, according to their estimate, consist of little over 2,031 square miles. Again, they make New Castle county, in their statement, contain 271,490 acres, and 619 square miles. When 271,490, divided by 640, would only bring 420, or make that county 195 square miles less by their report in acres than by their report in miles. The American Encyclopædia gives the area of the State at 2,160 square

[1] Murderkill was divided into two hundreds by Act of Legislature of March 20, 1867. Kenton hundred was formed from parts of Little Creek and Duck Creek hundred, by Act of Legislature, February 3, 1869.

miles. Huffington, of the Delaware Register, estimates it at 2,070. Mitchell's Atlas makes it 2,120 square miles. So far there has been no official survey of the State to verify what it contains.

Each of the counties take up the whole breadth of the State. New Castle and Kent are divided from each other by Duck Creek, and a line running from its northern branch about due west to the Maryland line. Kent is divided from Sussex by the Mispillion Creek and the Tan Trough Branch, one of its tributaries; thence southwesterly to a small branch of the Nanticoke, down this branch to the southward end of a beaver-dam, and thence by a line due west to the State line.

The accounts of most of the natural features of our State will be given under the head of its geology in the succeeding chapter. But as little mention of its rivers, streams and bays have there been made, and as it is necessary to know them, to comprehend the events hereafter to be related, a slight sketch is accordingly given of them and the other geographical features of the State.

The extreme upper portion of the State of Delaware (as will be found hereafter stated in our description of its geology) is composed of a mass of beautifully rounded hills, nowhere more than five hundred feet in height, situated on a sub-stratum of rock. Below the White Clay Creek, a distance of about seven miles from our quadrantal boundary, the land becomes level, the rock generally ceases, and a low sandy ridge, nowhere more than seventy feet high, passes through the State. This table land abounds in swamps, in which most of the rivers and streams of the State have their source. At the southern border of the State is a great morass, called the Cypress Swamp, about twelve miles in length, the whole of which is a high level basin. It contains nearly fifty thousand acres. About one half of this great swamp lays in Sussex county, Delaware; the other half in Maryland. It contains a great variety of trees and plants, mostly cypress trees, (called by the residents cedars,) and an immense quantity of huckleberry bushes, and is infested with wild animals. The deer and the bear, it is alleged, yet remain there. Below its surface are found immense trunks of cedar trees, the remains of giants of the forest long since gone. The residents of the locality probe through the morass with rods, to find where they are situated, and then raise them, and turn them into shingles for market. This whole swamp can be readily drained and made good land. The soil of the State is fertile. It has long been celebrated for its wheat. All the small fruits that grow in the temperate zone flourish here. It appears to be the natural home of the peach. Cotton was formerly grown in Sussex county. The noblest forests of white and black oak, yellow pine, cypress or cedar, tulip poplar, Spanish oak, gum, and other magnificent trees still exist in the State, although they are being rapidly cleared away. Its white oak, it is conceded, is the finest in the United States. Trees of this description in the Blackbird, Kenton, and other forests of the

State, are often three feet in diameter across the stump, and from forty to sixty feet in height. Logs are often sawed from them of thirty feet long. The black oak produces the best quercitron bark in the Union, and it brings a higher price than any other in the Liverpool market. It is ground principally at Milford and Smyrna. The Delaware Spanish oak furnishes the best known bark for tanning, and its bark brings a higher price, both in the foreign and domestic market, than any other. The forests of pine are principally in Sussex county. Sumach, which is worth from $50 to $60 a ton, grows wild in immense quantities. All the ordinary game birds, such as the snipe, the partridge, the old field plover, (a bird a little larger than a partridge,) abound in the State. The mocking-bird, rarely, if ever, observed north of our boundary, can be seen in numbers in Kent and Sussex, and the lower part of New Castle counties.

Its principal rivers, streams, and harbors are, first, the Delaware, which, for twenty-four miles from our northern boundary, is a part of our State. It is so well known as not to necessitate any description. Naaman's Creek, which flows into the Delaware about a mile from our northern border, is the most northern stream in the State. The Shelpot Creek, which flows into the Brandywine a short distance before it joins Christiana. The mouth of this stream is now dammed up. The Brandywine, which rises in Pennsylvania, and flowing through the State, dividing Brandywine from Christiana hundred, enters the Christiana within the limits of the city of Wilmington, about a mile and a half from where that river enters the Delaware. The Brandywine is navigable for about two miles from its mouth for sloops and schooners. From the termination of its navigation to the Pennsylvania line, it is mostly rocky, with several falls, which affords magnificent water power, from the city of Wilmington to the Pennsylvania border. Its banks are lined on both sides with mills and factories. The Christiana, which flows through the State in a northeast course from Maryland, and empties into the Delaware at Wilmington. This river is of sufficient depth to be navigated by vessels drawing 14 feet to the city of Wilmington, and sloops to the village of Christiana, about ten miles further. Red Clay Creek, Mill Creek, and Bear Creek, are large streams flowing into the White Clay Creek (a confluent of the Christiana) from the northern hundreds. They were once navigable, but are now valuable, mainly, for their water power. Red Lion Creek, formerly navigable, is now dammed up. St. George's Creek is now turned into the Delaware and Chesapeake Canal, except a portion of it which empties into the bay partially through a new channel. St. Augustine's and Silver Run are small creeks which flow into the bay in St-George's hundred, below Reedy Island. Appoquinimink is an important creek, and is navigable for sloops to Odessa, about seven miles from its mouth, and for steamboats to Thomas' Landing, about two miles and a half from its mouth. Blackbird is a navigable creek, which flows due east and north until it empties in the bay. Duck Creek, is an important creek, which divides New Castle from Kent county. It is navigable for seven or eight miles to within a mile of the important town of Smyrna, for whose exports it is the outlet. It is also navigable for several miles for steamboats, having water from twelve to fourteen feet to Hay Point Landing. It flows into the bay north of Bombay Hook, through a made channel called the Thoroughfare. Little Duck Creek is a creek navigable for sloops for several miles from its mouth to the town of Leipsic, the commerce of which it bears upon its waters. Dona River is a small river, the head of which connects with Little Duck Creek. For about three miles it flows with a broad channel, and enters the bay below Little Bombay Hook Island. This river and Little Duck Creek forms Kent Island, a large marshy island several miles in extent. It is navigable for steamboats for about two miles to Dona Landing. It was the place originally intended for the terminus of the Delaware Railroad. About two miles before Dona River reaches the bay, it is divided into two channels, one of which, (called Mahon River,) tearing itself a passage through the yielding marsh, and flowing southerly for about four miles, forms itself into a bay or harbor, and enters the Delaware. In conjunction with Dona, it forms Kelley's Island, another large marshy island, in which there are several ponds. This bay or harbor is called Port Mahon, and is esteemed the best harbor for coasters on the Delaware. Little Creek is a small creek, navigable for about three miles from its mouth for sloops and small schooners. It is the channel for the commerce for the town of Dover, the capital of the State, which hauls its exports four miles to a place upon it, called Little Creek Landing. It flows into the Delaware about a mile below the mouth of Port Mahon. Jones' Creek, which runs back of the town of Dover, is navigable for small sloops and schooners to Forrest Landing, about nine miles from its mouth. Forrest Landing is where the produce of the town of Camden is shipped. This creek is about twenty miles long, and flows in a southeasterly direction until it reaches the bay. Murderkill is a navigable creek, which flows in a northeast direction until it enters the bay about a mile below the mouth of Jones' Creek. Sloops and schooners can go up to the town of Frederica about ten or twelve miles from its mouth. Mispillion is a large creek, upon which the town of Milford is situated. It is navigable to Milford for large sloops and schooners, and steamboats have ascended it. It is the outlet for the commerce of that town. Mispillion is also the boundary between the counties of Kent and Sussex. Cedar Creek is a small navigable creek, flowing into the Delaware. It has also an artificial outlet into the Mispillion. Draper's, Slaughter's and Primehook Creeks are small unimportant streams flowing into the bay at various distances between the mouths of Mispillion and Broadkiln and Lewes Creeks. Broadkiln Creek is a stream navigable for sloops and schooners to the town of Milton, a distance of about twelve miles from its mouth. It flows in an easterly course, and enters the estuary of Lewes Creek about two miles from its junction with the Delaware Bay. Lewes Creek, from where it flows to the Delaware, to the town of Lewes, a distance of about six miles, is separated from the bay by Cape Lewes, a cape about six miles long, and varying from an eighth to three quarters of a mile in width. Large coasters used to sail up the creek, but it is now only navigable for boats. Its navigation was destroyed by what is known as the "Great Storm." The waters of the bay washed over the narrow cape, and filled the bed of the stream with sand. The navigation of Lewes is now through a canal from a small creek, called Canary Creek, to Mill Creek, and from there to Broadkiln Creek. Two small creeks, called Wolfe Creek and Old Creek, flow into it from the neighborhood of Lewes.

Rehoboth Bay and Indian River Bay are two large shallow bays, which are separated from the Atlantic Ocean by a narrow ridge of sand, which is from a quarter to half a mile in width, and about nine miles in length. Through this ridge the waters of the two bays have united, and torn a passage, called Indian River Inlet. This inlet rarely ever contains more than three feet water, and after a great easterly storm its mouth is generally stopped up by sand washed into it from the workings of the ocean; after which the waters of the bays again tear themselves a passage, and wash the sand which has filled up the inlet into the ocean. These large bays, each, contain about twenty-five square miles of surface, and at their deepest part do not exceed five feet. Their general depth is from three to five feet. Rehoboth, the most northerly of these bays, is nearly square in shape, and extends parallel with the ocean, separated from it by the ridge before mentioned for about six miles. It is probably six miles long by about five broad. Love Creek, Middle Creek, Herring Creek, and Guinea Creek flow into it. All of them shallow streams. It is separated from Indian River Bay by a neck of land called Long Neck, and several marshy islands, (now called Burtons,) but in the old maps called Staten Islands, where it is mentioned as abounding in oysters and terrapins. It is at this time, however, too salt for oysters to live in, and consequently none can be found there. Indian River Bay is about eight miles long, and from two to four broad. It only fronts about three miles on the Atlantic, from which it is separated by the narrow ridge before described. It extends lengthwise nearly due west into the State. Indian River, on which is situated the town of Milsboro', a broad shallow stream flowing due east, and of which it is a continuation, enters it. Pepper Creek, Vine Creek, and White Creek, streams of no importance, (although making a great show in the map,) flow into it. On Pepper Creek is situated the town of Dagsboro, and on White Creek that of Frankford. The small depth of water at the mouth of Indian River Inlet creates the necessity of forcing the shallops over the bar by kedging. This causes a thumping of the vessel's keel on the sand, which drives the bilge water into the hold, and spoils much of the grain, which is the general cargo of these vessels.

Fresh Pond and Salt Pond are the names of two remarkable ponds in Baltimore hundred, situated on the Atlantic coast, a few miles south of Indian River Bay. Fresh Pond is about half a mile long by one or two hundred yards wide, and about twenty-five or thirty feet deep. It has no outlet, and apparently no streams flowing into it. It contains beautiful fresh water, and a few fish. The ridge of sand between it and the Atlantic is not more than an eighth of a mile wide. Great storms sometimes wash away a portion of this ridge, and let the salt water into the pond. But the ocean again forms the ridge, and restores things to the condition they were in before.

Salt Pond is another body of water about the size of Fresh Pond, and situated within about three miles to the south of it. It is probably one-half of a mile further from the ocean than Fresh Pond, and the Atlantic does not break through its banks and encroach on its waters, as in the case of Fresh Pond. It, like Fresh Pond, has no outlet. Its waters are very salt, far more so than those of the ocean from which it is separated by such a slight barrier. Indeed, it is so salt that no fish can live in it. Salt works were once erected on its banks, and a great deal of salt extracted from it. Salt is still manufactured from its waters by the citizens of the neighborhood for their own use.

Assawaman Bay is the last body of water in Delaware. The head of it is formed by Jefferson Creek, which flows into it from the north. It is a long shallow bay about seven miles long from one to one-half a mile broad, and from four to five feet deep, navigable only for boats. It is separated from the Atlantic by a long narrow ridge of land, from a mile to three quarters of a mile wide, called Fenwick's Island. It flows into St. Martin's Bay, in the State of Maryland, which is about fifteen miles long, and which has its outlet in the Atlantic through an inlet formed by the termination of Fenwick's Island and the Island of Assateague. After passing Fenwick's Island it takes the name of Sinepuxent Bay. Williams' Creek, a shallow stream, flows into Assawaman Bay. Fenwick's Island is not an island, but a long narrow cape and ridge of land, generally from half a mile to a mile wide, and about twenty-three miles long, covered with oak, scrub-oak and pine, about one-third in Delaware, and the other two-thirds in Maryland. It, with the islands of Assateague, Chincoteague, Wallop's, and other islands form a series of shallow sounds stretching from the southern boundary of our State to Cape Charles, at the head of the Chesapeake Bay. On this island, at the Atlantic, the southern boundary line of the State of Delaware commences. Hence the saying of Delawareans when they wish to express the utmost limits of the State, "from Naaman's Creek to Fenwick's Island," similar to the expression of the Israelites, of from Dan to Beersheba, or that of the Britons, from Land's End to John O'Groats.

This finishes our list of creeks and harbors on the Delaware River and Bay. This noble bay, which, with the Atlantic, forms our eastern boundary, abounds with fish and fowl. Outside of the capes, on the Atlantic, milletts are caught in immense quantities by angling. Within the Assawaman, Indian River, and Rehoboth Bays fish and terrapins abound; the fish are mostly rock, flounders, perch and eels. Around the Breakwater are caught immense quantities of black-fish and lobsters. Within the bay are taken the perch, the trout, the alewive, the sea crocus, (better known by the name of the spot,) the sheepshead, as well as the drum and the eel. Terrapins abound on the shore, and oysters are found in immense quantities in various portions of the bay, and in Mispillion, Broadkiln, Port Mahon, and other rivers and harbors of the State. Shad and herring are found in the bay, river, and all the principal streams. The man-nin-nose, a delicious shell fish, (shaped something like a clam, only with a soft shell,) is dug up from under the sand.[1] The kingcrab is cast up in untold numbers every tide by the waters of the bay, and although not fit for food, serves a valuable purpose by being fed to hogs, and ground up into what is now called "cancerine," and placed upon the land, where it has proved a most valuable fertilizer. The crocus, (or spot,) the sheepshead, and the drum are particularly plenty, and are by some thought to be peculiar to the Delaware. The crocus, or "spot," derives its name from a little black spot on each side of its head about as big as a five-cent piece. The sheepshead have a mouth and teeth exactly like a sheep, and are nearly as broad as long. The drum fish are caught principally on Mispillion (generally

[1] I do not know whether the name of this fish is spelt correctly. I have never seen it either written or printed, and never heard of it out of Delaware.

called by the residents Mushmellon) Flats, in the bay opposite the mouth of the Mispillion. They get their name from, when swimming under water, making a noise like a drum. They are caught with a hook and line, and often weigh twenty pounds. From their weight there is often difficulty in hauling them in. In addition to wild ducks and geese, the curlew and crane frequent the shores of our State. There are but few of the latter, however, to be seen.

Although several streams flowing into the Chesapeake Bay have their rise in Delaware, yet the only three of any importance are the Nanticoke, the Broad Creek and the Pokomoke. The Nanticoke is navigable for large schooners and steamboats to the important town of Seaford. The Broad Creek, to a village of a few houses, named Portsville, about three miles from the town of Laurel. The Broad Creek flows into the Nanticoke eight or nine miles below the town of Laurel, and about the same distance from the town of Seaford. The Pokomoke River, which flows through and past the Cypress Swamp in a southerly direction, is navigable for small vessels. The principal streams which take their rise on the ridge in this State, and flow into the Chesapeake, are the Back Creek, the Bohemia, and the Sassafras, in New Castle county. The Chester, the Choptank, and the Marshy Hope, in Kent county, and the Wicomico, in Sussex county. The Marshy Hope derives its principal importance from being deepened and arranged as a drain for the celebrated marsh of that name in Kent county. All the above mentioned streams expand when they reach Maryland into large and important rivers.

CHAPTER II.

Geology of the State—Its Northern Boundary Curved—Underlaid by Beds of Rocks north of Philadelphia, Wilmington and Baltimore Railroad—High Hills—Depth of Soil—Rocks composed of Gneiss, Feldspathian Rock, Limestone, Serpentine, Granite—Decomposition of Gneiss in Christiana and Mill Creek Hundreds—Depth of Deposits—Garnets—Granite fused into Rock—Magnetic Iron Ore in Christiana Hundred—Feldspathian (blue rock) from Wilmington to Naaman's Creek—Hardness of the Rock—Shellpot Hill composed of it—Limestone in Mill Creek and Christiana Hundreds, of excellent quality—Serpentine Rock, abounds in Feldspar—Spar Quarries—Asbestos—Chromic Iron—Hematic Iron Ore and Kaolin in Mill Creek Hundred—Rock overlayed by Deposits of Sand and Gravel—Soil of Upper Hundreds tenacious and heavy—Second, Tertiary and Recent Formations—Table Land of the State—The Ridge—Red Clay Formations—Clay under New Castle, Christiana Village, Red Lion Hundred—Iron Hill and Iron Ore—Red Clay Formation extends over Hundreds of New Castle, Red Lion and Pencader—Green Sand or Marl Formation extends over St. George's Hundred—Value in Agriculture—Analysis of Marl in Deep Cut of Canal, Organic Remains, Amber, Ammonite, Lignite, Tertiary—Yellow Clays of Appoquinimink—Blackbird Hill, the Levels and Ponds in them, Petrified Wood—Lower Part of Appoquinimink Loamy—Geology of Duck Creek Hundred, Rock—Soil of Kent County—Geology on Murderkill Creek—Shells, Clays and Sands of Kent—Bog Ore—Springs at Dover—Neck Lands of Kent—Marsh Lands of the Ridge in Kent—Soil Burns like Coal—Recent Formations—Clay and Sand of Sussex, Clay Predominates—Soil of Sussex County, Stiff Clays and Blowing Sands—Medium Texture in Northwest Fork Hundred and the Neck—Course of the Clay Bed—Iron Ore in Sussex—Sands of Sussex—Soil Blown away by the Winds—Sand Hills—River Deposits—Neck Lands of the State formed by them—Blue Mud—Salt Marshes of the State, can be embanked—Chalybeate Springs and Water in the City of Wilmington and over the whole State.

THE geological formation of the State is as follows: the surface of its northern or curved boundary to a line crossing it at a little north of the Philadelphia, Wilmington and Baltimore Railroad, and containing about one hundred square miles of territory, (comprising the hundreds of Brandywine, Christiana, Mill Creek, and part of White Clay Creek,) is composed of a comparatively confused mass of beautifully moulded hills, with a bold and rounded outline, always elevated, and often rising several hundred feet above tide water, and affording an outlet to the waters of rain, springs and creeks, through deeply cleft valleys, with rounded or abrupt rocky sides. These hills, however, nowhere exceed five hundred feet in height. The whole of this district is underlaid by continuous beds of primary rock,[1] which, in many cases, may be observed cropping out of the ground. The soil, though varying in particular localities, is generally uniform in its character, being argillaceous,[2] and mingled with a greater or less quantity of gravel, while an alluvial[3] deposit of a similar character covers the rocks to the variable depth of from one to sixty feet, and perhaps still more. Of these rocks there are five different kinds, called by geologists gneiss,[4] feldspathic[5] rock, limestone and serpen-

[1] Rock supposed to be first formed.

[2] Composed of clay.

[3] Composed of matter deposited by the water.

[4] A stratified primary rock, generally composed of quartz, feldspar and mica. Mica is a transparent glistening mineral.

[5] A silicious or flinty rock. The constituent parts of which are silica, alumina and potash.

tine.[1] The remainder is of a granitic character. Nearly the whole (or $\frac{99}{100}$) are gneiss and feldspathic rock, of which the former (gneiss) composes about three-fourths of this portion of our State. The mica is more prevalent in it than its other constituents. This gneiss is of various degrees of hardness, in some instances it can be cut by a knife, and then again the steel itself is abraded by it. This decomposition of the gneiss may be observed chiefly in the southern portion of Christiana and Mill Creek hundreds, and along the Newport Pike, in the neighborhood of Newport. In some places the deposits above the rocks of this decayed stone are from twenty to thirty feet, and how much more has not been determined, although it is believed the solid rock does not lay far beneath. In the northern portions of Brandywine, Christiana, and Mill Creek hundreds this stone becomes hard in its character. In many places, amongst the northwest portions, the two former hundreds, the common red garnet[2] is diffused amongst it. On the Kennett Pike, about five miles from Wilmington, the quartz has the appearance of having been ejected or fused in a liquid state into this rock. Magnetic iron ore is occasionally found in the northwest portion of Christiana hundred.

The feldspathic rock, better known by the common name of blue rock, occurs chiefly along the Delaware,

[1] A species of rock or mineral crystallized and massive, and also fibrous and foliated or leaved, and composed chiefly of hydrous silicate of magnesia.

[2] A mineral or gem of a red color.

from Naaman's Creek to Wilmington, and from thence may be observed at intervals to the western State line, being indicated by projecting masses of the solid bed, or by huge boulders, upon whose sharpened outline the atmosphere appears to have had no effect. This rock extends over one-fourth of the primary portion of our State, the gneiss occupying nearly the other three-fourths. It is composed generally of crystalline masses of smoky feldspar and quartz, with occasional plates of mica, and more rarely veins abounding in hornblende.[1] The feldspar is the most abundant, sometimes constituting the entire rock. This rock is found also along the Brandywine and the Concord Turnpike. Shellpot Hill is supposed to be entirely composed of it. It abounds on the top of the hill skirting the Brandywine near Riddle's Factories, although there it is of so light a color as hardly to merit the name of blue. In blasting, enormous masses of this rock are frequently heaved from the solid bed, and are capable of being split in wedges in any required direction.

Limestone, although occupying an unimportant extent of country, in comparison with the gniess and feldspathic rock, occurs in great abundance near Pike's Creek, and near Hockessin Meeting House, in Mill Creek hundred, where it is extensively quarried. It also occurs about two miles west of Centreville, and at Smith's Bridge, in Christiana hundred, and where the Brandywine enters the State. It is generally

[1] Having the ingredients of silica, magnesia, lime and iron.

pure marble, essentially composed of lime, magnesia, and carbonic acid. It lays in heavy beds. This lime is unequalled for agricultural and building purposes.

Serpentine occurs about six miles northwest from Wilmington, where it exists in its greatest extent on a ridge of about one mile in length, with a breadth of about half a mile. At one place it rises in a green rock, abruptly from a meadow near Green Hill School-house, in Mill Creek hundred. It is traversed by a granitic vein so rich in pure feldspar as to originate the "Dixons Spar Quarries," in order for its employment in the manufacture of porcelain. Asbestos[1] can be found there in large quantities, as well as other minerals incidental to it, amongst which is chromic iron, a mineral of some value. But it has not been detected in any quantity, although diligently sought after. Another body of serpentine exists near the State line, and where the Brandywine enters the State, generally of a light green color, and containing hematitic iron ore of a good quality. It is this ore that gives it its green color.[2]

The granite found in Delaware is principally in a

[1] This is a fire-proof mineral; can be woven in connection with cotton, tow or other textile fabrics, and fire-proof garments made from it. It is a common practice to cleanse garments of asbestos by throwing them in the fire.

[2] Grace Church, in the City of Wilmington, the most magnificent Methodist Church in the United States, is built of serpentine. It was obtained near Chadd's Ford, the site of the Brandywine battle-ground, about three miles beyond the Delaware line. It is part, however, of the same bed with that in Delaware.

vein traversing the serpentine rock. It is characterized by its abundant contents of very pure white feldspar, unusually free from oxyd of iron, a circumstance from which it derives its principal value. It also contains numerous minerals. Indeed, the serpentine and granite together, offer the finest mineral locality contained in the State of Delaware. When feldspar undergoes decomposition it forms a species of clay known under the name of kaolin, which possesses great value in the manufacture of china ware and porcelain, especially, when free from oxyd of iron. This substance has been found in a few localities, but being deposited in small brooks, it is too limited to demand attention. A large deposit exists just across the Pennsylvania line, from Mill Creek hundred.

Overlying the primary rocks of our State is a diluvial deposit of clay, sand and gravel, arising from the uneven surface of the rocks, and amounting at times to at least sixty feet in depth. On the south-eastern edge of this rocky region, it consists of a dark red clay, embodying in places a large proportion of gravel, as may be seen more strikingly in the cuttings of nearly all the roads through that region, in the neighborhood of Wilmington. This deposit of clay has a necessary influence in rendering the soil of the upper hundred tenacious and heavy.

South of the Christiana and White Clay Creek the geology of the State undergoes a change, and instead of rocky or primary formation, we have what is called upper secondary, tertiary, and recent formation. It presents a comparatively level country, or table land, generally sloping east and west towards the Delaware and Chesapeake Bays from an elevated strip of land several miles in breadth, on which the streams flowing east and west take their rise in swamps and morasses, and cut their channels through the yielding soil. The name of water-shed or dividing ridge is applied to this narrow track, which extends through the whole length of the State. From the boundary of the primary or rocky region to the southern boundary of New Castle county is a series of clays, sands and gravels, which are called upper secondary formations. But all the tracts bordering on the Delaware, from New Castle to Sussex, has an argillaceous or clayey soil. This upper secondary formation is called by Booth,[1] the geologist, from whom we extract our information, the

RED CLAY FORMATION,

from their being composed of a series of clays, in which the red color predominates, although inconsiderable veins of white are mixed with it. The red clay is often covered with heavy beds of sand and gravel, although it may often be seen cropping out of the ground, and again may be observed where cuttings have been made for the roads. It underlies the town of New Castle, where beds of it have been penetrated at one instance to the depth of seventy feet, and in

[1] James C. Booth, who was appointed for that purpose by the Legislature, made a geological examination of the State in 1837 and 1838.

another case of one hundred and eighty feet. At one place below this town, on the Delaware shore, however, the white clay predominates, where it is found in great abundance and purity, and has been used for the manufacture of crucibles for making glass and other valuable articles for near seventy years. The neighborhood of Christiana village affords the best place for studying this red clay formation, which presents excellent sections in the banks of the creek, and on the road cuttings, the upper portion frequently lying from sixty to eighty feet above tide-water. The red clay may also be observed in numerous places in New Castle, Pencader, and Red Lion hundreds. Along Red Lion Creek and its tributaries it crops out of the sides of the hills, and generally constitutes the beds of the streams. Iron pyrites[1] are sometimes abundantly contained in it, which accounts for the frequent occurrence of iron stone and ferruginous bands. From Newark to Summit Bridge, on the dividing ridge, the soil is argillaceous, composed of white and yellow clays, with a bed of gravel and sand below, which rests on another bed of clay. From the track of what used to be the Frenchtown Railroad to the Summit Bridge the geology is little else than soil and gravel. In the vicinity of White Clay Creek there are several outlying spurs of considerable height, of which Iron Hill is the most important. This elevation, rising abruptly from, and traversing far above the plain, consists of clays, sand and gravel,

[1] These pyrites are of a yellow color, shine like, and have the appearance, of gold. They are frequently called "fool's gold."

and derives its name from the abundance of iron, stone and ferruginous[1] quartz scattered over its flanks, the latter of which was probably at one time of good quality, but through exposure to atmospheric agents has been rendered valueless. An excavation has been made on the summit for the extraction of iron ore to the depth of forty to fifty feet, which enables us to estimate the character of this singular hill. The mass of earth is highly argillaceous loam, interspersed with large and frequent masses of yellow, ochry clay, some of which are remarkable for fineness of texture, not unlike lithomarge,[2] and consists of white, yellow, red, and dark blue clays in detached spots. Nodules of iron ore are abundantly distributed through the whole formation. Large quantities of this ore has been exported. The depth of the red clay formation is estimated at 250 feet.

Towards the southern border of the red clay formation, which extends over the hundreds of New Castle, Red Lion and Pencader, and which consequently possess a soil of a heavier nature than usual, a deposit of yellow sand begins to appear, in elevated situations, and becomes gradually thicker as the red clay sinks below it, until we approach the Delaware and Chesapeake Canal, where it takes up, and includes in it, the celebrated green sand or marl, which has added so much to the fertility of the lands in the neighborhood. This has been denominated the

[1] Partaking of the nature of iron.

[2] A mineral used for drying paint.

GREEN SAND OR MARL FORMATION.

These two sands in combination occupy the whole of St. George's hundred. The yellow usually preponderates over the other, sometimes presenting bluffs fifty feet in height, whilst the green sand rarely exceeds thirty feet in thickness. There appears to be two principal deposits of green sand, the upper and lower, which rarely unite to form one stratum, and are often separated by twenty or thirty feet of yellow sand. The lower stratum is chiefly confined to the canal, whilst the upper, although visible at the Deep Cut in the canal, first assumes importance several miles to the southward. Both deposits derive their character from a green substance with which they abound, and which being in the form of small grains, received the name of green sand, but this granular form being the only property in common with ordinary silicious sand, the two should not be confounded together; for whilst the latter contains principally one ingredient, silica or flint, the green sand is composed of five or six, amongst which are potassa and lime, two substances of the highest value in agriculture. The soil on the neck lands and dividing ridge of St. George's hundred is rather argillaceous and heavy, but through the rest of the hundred the yellow sand, rising to the surface, assumes the character of a loam, that can scarcely be excelled for the well proportioned mixture of fine sand and clay, and proves itself capable of the highest degree of improvement. An analysis of the two descriptions of marl are as follows, viz.: in one, carbonate of lime, 20; green sand, 38; silicious sand,

32; clay, 10; total, 100. In the other, silica, 58·10; potassa, 7·55; protoxide of iron, 22·13; alumina, 5·14; lime, a trace water, 8·22; total 101·13. The quantities of these constituents, however, differ in various degrees in the different localities. The following shows the strata of the State at one place, through the deep cut of the canal, at about one-quarter of a mile west of Summit Bridge, viz.: soil, ferruginous gravel and sand, 9 feet; black tenacious clay, 7 feet; ferruginous brown sand and clay abounding in ammonites,[1] baculites,[2] lignites[3], and amber[4] for 23 feet. Blue micaceous sand and clay, with same organic remains as above for 11 feet. Ferruginous sand and clay of a dull green color for 6½ feet. White silicious sand and lignite abundant for 5 feet. Total, 61½ feet. This reaches the bottom of the canal, and with two others, within a quarter of a mile of it, are the deepest diggings made into the soil of Delaware. Exactly at the Summit Bridge the depth of the cut to the bottom of the canal is, however, 74 feet, and they bored three feet below the bed of the canal. In this they found iron crusts, and numerous organic remains which extended as far as the boring. Nu-

[1] The shell of an extinct shell fish like a coiled snake, called snake stone. Called so for ornaments like it being on the chair of the statue of Jupiter Ammon.

[2] A shell similar to the ammonite.

[3] Petrified wood.

[4] A beautiful gum, belonging to trees long extinct, used for mouth pieces of pipes, &c. A piece of amber was also found thirty or forty years ago on the farm of Mr. S. Higgins, on the State Road near the canal. This is the only place in which amber has been found in this State, save in the deep cut of the canal.

merous springs flow into the canal from this deep cut, holding in solution so great a quantity of sulphate of iron (copperas) as to kill all the fish and the barnacles on the bottom of vessels passing through it. All the springs nearly of the deep cut give evidence of copperas in their taste, and make irony deposits at their point of issue. Iron pyrites constantly abound through the sands and clays.

Between St. George's hundred and the lower part of Kent county the geology of the State again undergoes a change. Between these points exists a series of beds of clays and sands, comprising two narrow belts abounding in organic remains, which are different from upper secondary, and therefore the appellation of

TERTIARY

has been applied. When the green sand or marl reaches Appoquinimink hundred it descends below a yellowish clay or loam, which underlies this hundred, from which it is separated by a ferruginous sandstone, sometimes six feet in thickness. At these points the clay is not more than fifteen feet thick, but when it reaches Blackbird it constitutes a hill thirty feet in height, and occasionally alternates with deposits of yellowish sand. This contains a large quantity of silicious sand, and in many localities rises nearly, or quite, to the surface, and imparts to the soil of a large portion of this region a considerable degree of tenacity, and consequently difficulty of working. The neck lands, however, as well as the western part of this hundred, offer a soil of superior quality. The latter, known as the Levels, has long been celebrated for fertility. The soil is a fine loam, and capable of retaining more moisture in consequence of the yellow clay at no great distance below the surface. Indeed, the proximity of this clay, combined with the level character of the country, proves an annoyance to farmers from ponds of water forming in the fields, which often lie long enough to injure the crops. This yellow clay abounds in fragments of petrified wood, belonging to an ancient species of pine, the only specimen of organic remains hitherto detected in it, one of the best localities of which is on the road from Odessa to Blackbird, between half and three-quarters of a mile from the former place, where it is profusely distributed in the gravel, and plowed up in the fields, in masses sometimes weighing thirty pounds. From the nature of the clay it is supposed that they have been transported in the state of wood to this place, when it was below the surface of the water, imbedded in the fine mud, and there have been converted into stone, the vegetable matter having been replaced by silica and alumina. Similar petrifactions occur at intervals lower down the State. In some of the streams of this hundred are found conglomerated masses of gravel cemented by oxyd of iron. In the lower part of this hundred are found at times the light loam *and sands of Kent county.*[1]

[1] In digging a well some years since on the farm of Mr. Benjamin Dennis, near Townsend, in this hundred, the laborers came to a hard bed of sand of the consistency of stone, which could be lighted by a candle or match, and would then burn brilliantly.

In the northern part of Kent county, on Old Duck Creek, about four miles from Smyrna, and other places in that vicinity, a strong crust is observed topping out of blue clay, containing abundant casts and impressions of shells. Crossing the dividing ridge, at the head of the Choptank, near the mill, there is a blue clay resembling that of Old Duck Creek, and which is supposed to be a continuation of it, excepting in the absence of shell impressions. From these clay deposits emanate a quantity of copperas. Near that branch of the same creek, lying immediately south of Smyrna, are found large masses of silicious rock, which, from its extreme hardness and toughness, could scarcely be supposed to have had its birth in this region of soft clays and light sands, had it not been found in place in one locality. So numerous and large were the blocks on a farm belonging to the late Mr. Cloak, on the State Road, on the south side of the branch, that it was found necessary to sink them in order to the better cultivation of the land. They consist of coarse sand and gravel cemented by silicious matter, containing frequent casts of shells. Large blocks of the same substance were found on other farms in the vicinity. Silified shells are found abundantly on the Kenton Road, five miles from Smyrna. Although the soils of the northern part of Kent county are very valuable, yet we may distinguish three things which are most prevalent, viz.: those of the neck lands of a heavy character, but supposed to be the most fertile in the State; those of the dividing ridge, consisting of very heavy bottoms, not always productive, with occasional light sand hills; and lastly, those intermediate between them in position, and occupying a much greater space. They are light loams easy of culture, which, by the energy of the farmers, are being brought up to the highest degree of fertility.

Crossing the middle sections of Kent the tertiary is found more fully developed on the streams of Murderkill Creek. The lowest stratum is visible at the head waters of this creek. It is blue clay, closely resembling that at Duck Creek. It abounds in impressions of shells in a soft clayey state, and is separated from an overlying white sandy bed by a hard ferruginous crust, containing similar organic remains. This white sandy bed is a half-hardened mixture of sand and clay, consisting almost wholly of shell casts, containing but in a solitary instance a trace of lime, which was a shell found unaltered. Twenty-five feet is the greatest ascertained depth of the tertiary. Its clearest indications are observed at Spring Mills, on a fork of the Murderkill of that name, near Frederica.

In addition to the tertiary there are in Kent County, for a distance of about twenty miles, a series of beds of clays and sands, the lowest of which is clay, observable on nearly all the streams, varying in color and texture in the same locality, but generally of a yellowish shade, and of medium fatness or richness, and the upper consisting of ferruginous sands or gravel. Indications of these formations may be seen on the branches of Little Creek; at Dover, where the sandy nature of the upper beds, and the upper level of the clay, are indicated by the numerous springs of excellent water which issue from the foot of nearly every declivity. Below the entrance of the Tydbury Branch into Jones' Creek, at Forrest Landing, where the clay rises some ten feet above tide-water. On one of the head branches of Jones' Creek, near Rashe's Cross-roads, beds of solid iron bog ore, of four-tenths inches in thickness, was found for the distance of a mile. From the similarity of soils throughout a great part of Kent county, in the same relative situation they may be classed as was done with those in New Castle county, viz.: with the exception of the marshes bordering on the bay, all that low land, known as the neck lands, is of a heavy argillaceous character, and naturally remarkably fertile, although some of them have been subject to more than a century of excessive tillage. As we rise, the country to the westward, and meet the tertiary deposits, the soil becomes more loamy, corresponding with the subjacent deposits, and as these become covered by loose sand towards the ridge, the surface necessarily partakes of the same character. Proceeding from north to south through this middle section, the amount of loose sand increases in depth and breadth, so that much of the land in the lower part of Kent county has a sandy soil. On the ridge we find the same alternations of light sand hills and heavy clay bottoms, which was noticed in New Castle county, but which in Kent are more strongly contrasted. In addition, however, to these is a vegetable soil too remarkable to be passed over by a simple notice, viz.: the marsh lands, situated in the southern and western part of Kent, and on the ridge, in Sussex. They are situated on the branches of the several streams, which, having their source in the Delaware, usually flow towards the Chesapeake, and which, originating from rains and springs in the midst of extensive forests, on a broad and very flat surface, with a clayey substratum impervious to water, and becoming clogged and dammed up by fallen trees, leaves, and other brushwood, naturally expand into broad basins termed marshes. The luxuriant growth of trees, shrubs, and smaller plants, and their constant dilapidation and decay in the shallow waters of the sluggish streams during the lapse of ages, has generated a black vegetable mould, averaging three feet in depth, being rarely less than six inches, and sometimes exceeding six feet, being composed throughout of the same materials. It was not until the close of the past or the present century that effectual means were resorted to for recovering this land from almost constant inundation, since which time nearly all the great marshes have been drained by the excavation of ditches, or, more properly, canals, in the natural bed of the stream, and a large amount of the most fertile soil of the State brought under cultivation. One of these great ditches increases from twelve to twenty-four feet in width from its source to its mouth, a distance of nine miles, and throws off a sufficient quantity of water in spring to float a moderate sized vessel. The Colbreth, Cow, Herrington, and Tappaannah marshes, on the west of Kent county, are the main feeders of the

Choptank, and Marsh Hope, in the south, forms a main source of the Northwest Fork River. Herontown Bog, another of the great marshes of Kent, was drained through the energy of the late ex-Governor Tharp and Alexander Johnson, Esq. This land, which had for ages been the roost of herons, when cultivated produced from ninety to one hundred bushels of corn to the acre. The principal and several minor branches of the Nanticoke have also been subject to drainage, besides many smaller streams in Sussex. When all the water courses shall have been confined in a similar manner, in their proper channels, which, from the constant improvements going on in the State, will not be long, a very large amount of fertile soil will be brought under the plow, and the noxious exhalations of marshy lands will cease to produce disease, to which their inundated state renders them subject. The soil of these marshes, when drained, are rarely so light and spongy as not to admit of the growing of grain after a little cultivation. The quantity of organic matter in them is so great that during a dry season the soil which was accidentally fired continued to burn like coal, and was only extinguished by rain. The remains of such fires have been observed in several instances, when the carbonaceous matter having been burned out, left the earthy constituents converted into a substance resembling brick by the heat of the fire.

To the southward of the lower tertiary, and as far as the southern limits of this State, containing the hundreds of Mispillion and Milford, in Kent county, and the whole county of Sussex, the geology of Delaware undergoes another change, and what are called

RECENT FORMATIONS,

or land supposed to be recently formed by nature, occur. They consist of a deposit of clay and sand, with a preponderance of the former, to which no date can be assigned, except in a few instances, in consequence of the absence of organic remains, and the impossibility of drawing any conclusion relative to their age from their mineral character. Thus the blue clay on Murderkill Creek could not be distinguished from similar clay on the shores of the bay, which is quite recent in its origin, except by comparing the ordinary bay shells of the latter with tertiary shells in the former. In few cases where shell beds have been found, there are no indications of tertiary fossils, the shells being referable only to the same genera and species which now inhabit the waters of the bay. The surface of the country of this lower part of Kent, and county of Sussex, is much more level than the other portions of the State, and less scooped out in ravines. The soil is also more variable, offering the two extremes of stiff clays and blowing sands. But in several instances, as in Northwest Fork hundred, and on the neck lands, it is of medium texture, and endowed with superior fertility. The greatest thickness of the clays is forty feet. The lowest stratum is a yellowish clay, at times of a light lead color, alternating with thin seams of sand, and superimposed by yellowish and nearly white sand of very variable thickness, amounting at times to twenty feet. On the Mispillion Creek, in the vicinity of Milford, and to the west of it, the uppermost stratum is of loose sand, with an occasional argillaceous bed, below which is a heavy bed of clay, extending below the water level. On Mill Branch, about a mile from Milford, the upper stratum is a loose yellow sand, which is underlaid and stratified by a white clay, below this is a fat yellow loam reposing on white sand. On Cedar Creek below Milford, the clay lies at a lower elevation, whilst the superimposed sand is of considerable thickness. From the abundance of superficial loose sand, the soil of this region derives its character; but where this has been partially removed there are basins, with a substratum of clay, which, being impervious to water, constitute ponds, that are sometimes a convenience, but oftener an annoyance to the farmer. At Milton the clay rises to the height of forty feet above tide-water. Proceeding westward to Lewes the same beds of clay present themselves at Cool Spring. The same clay crops out at the beach, three miles south of Cape Henlopen, where it has been uncovered by the gradual encroachment of the ocean, and although no traces are found of it at the head of Cypress Swamp, the supposition is that this clay underlies the whole country between Indian River and the Nanticoke, constituting the bases of the swamp, for it is well developed in the vicinity of Laurel, and on nearly all the branches of the Nanticoke, rising from five to twenty feet above tide-water. Under the town of Seaford are a series of clays of yellow and white colors, with occasional seams of sand, and intervening crusts of iron stone, the whole rising fifteen or twenty feet to the sandy soil. Small nests of shells are found in digging on the banks of the Nanticoke, between Seaford and Concord, near Cannon's Ferry, and at various other places. Iron ore of the bog variety is found in several parts of Sussex county. Amongst them at Little Creek, and on Broad Creek, about two miles east from Laurel; a few miles northwest of Georgetown, on Deep Creek; on Green's Branch, eleven miles west of Millsboro'; on Burton's Branch, one mile from the same town. At one time there were a number of forges and furnaces that manufactured this ore into iron, and the county of Sussex was mostly supplied with that material from its own works. A great deal of iron ore was also exported from Sussex. But the furnaces are all now idle, and but little, if any, of this material is now sold abroad.

To the traveller who for the first time passes through Sussex county, the formations would appear to consist almost entirely of loose white and yellow sands, but a more thorough investigation shows the fallacy of such a conclusion, and proves that in reference to geological deposits the clayey greatly predominates over the sandy, forming the substratum of the whole county, but that the latter, overlying and capping the clay over a large proportion of the surface, communicates the well known sandy character to the soil. These two upper sands probably cover one-half or two-thirds of the county, are of variable thickness, sometimes yellowish and more tenacious, at other times nearly white, and so loose as to be readily transported by the winds. This sand is of so loose a nature, that when the sod has been removed in an exposed situation, the action of the wind roots it out to the depth of several feet, distributing it around the surrounding soil, or heaping it against a bush fence or other obstruction. This sand is sometimes blown into steep hills, in several parts of the county. These sand hills, however, must not be confounded with the hills partly of sand and gravel which exist and which is due to the action of the waves, when the State was covered by water, and which was formed in a similar manner to the bars now formed in the bay. The most striking of these hills is one lying to the south of Milton, and between Georgetown and Lewes. It is a ridge of variable breadth, and not more than fifty feet in height, apparently in a north-west and south-east direction, composed of fine gravel and sand, and a sufficient mixture of clay to render it compact.

There is another branch of the geology of Delaware which comes under the name of

RIVER DEPOSITS,

by which the lands bordering on river and bay, called our neck lands, have been formed. The Delaware has been engaged for ages in transporting sand and clay from the northward, by means of which the sand-banks and shoals of the bay have been raised, and the heavy soils on the neck lands deposited. A large number of these shoals are, in all probability, remnants of the land which united Delaware and New Jersey prior to the wearing away of its channel by the river, but they have been increased, and many of them entirely formed by the detritus brought down by the river. A boring made on the island, on which Fort Delaware is situated, by Booth, the geologist, proved it to have been wholly formed by the river deposits. This deposit is generally known by the name of *blue mud*, and it is this blue mud which gives the neck lands of our State their great fertility. These neck lands are those tracks which border on the Delaware from New Castle to Sussex. The most celebrated among these are Raymond, Little Creek, Prime Hook, and Slaughter Necks. Raymond's Neck is supposed by many to contain the best land in the State. It is supposed that the noblest forest in the State exists at Prime Hook Neck, in Sussex county, consisting of tulip poplar, black walnut, and black oak, remarkable for their enormous size and flourishing condition.

From the upper part of the State to its southern boundary, on the Delaware River and Bay, and the sea-coast, are skirted by marshes of varying breadth, (better known by the name of salt marshes,) sometimes exceeding two miles, subject at times to inundations, consisting of flat and dark colored vegetable mould, and clothed with a luxuriant growth of reeds and grasses. These marshes are supposed to contain between one and two hundred thousand acres. This marsh, it is supposed, could be embanked, and the land brought into cultivation. In the upper portions of the State this has been successfully done. Undoubtedly the whole of this land will finally be recovered from the water, as the substrata of the marshes from the town of New Castle to the southern boundary of our State are similar.

We shall close our description of the State by stating that in many parts are chalybeate springs of excellent qualities, and great medical virtues, amongst them, are the celebrated Brandywine Springs, about three miles from Wilmington; others near Brackenville, in Mill Creek hundred; on the Brandywine Creek, about a quarter of a mile from Wilmington; in Duck Creek hundred, not far from Smyrna; at Spring Branch, near the Town of Frederica; near the Town of Laurel, in Sussex, and various other places too numerous to mention, in all the counties of the State. In nearly every quarter of the City of Wilmington, chalybeate water is reached by digging wells. A pump of this water used to be in operation near the corner of Sixth and Spruce Streets, now covered by buildings. Another is yet used within the city limits, on the Newport Pike.

CHAPTER III.

Description of the State before the Arrival of the White Man—Fruit and Corn Grow Wild—Diminishing of the Streams—Disappearance of Christina Harbor—Diminished width of the Christiana—Cherry Island Marsh—Overflowing of the Marshes—Diminishing of Clement's Run—Shellpot Creek—Encroachment of Wilmington on the Christiana—Bars formed at Blackbird and Mispillion Creeks—Disappearance of Stone Wharf Creek and Synapuxent Inlet—Dimunition of Lewes' Creek—Loss of its Trade—Description of Lewes' Creek from a Manuscript in the British Museum—Encroachments of the River, Bay and Ocean on the State—Destruction of Graveyard at New Castle—Port Mahon, Duck Creek, Dona River, Bombay Hook Island, the Thoroughfare—Destruction of the Sand Hills on the Atlantic—Hen and Chicken Shoal and Island—Increase of Beach near the Breakwater—Extension of Cape Henlopen—Abundance of Fish—Law passed forbidding the Erection of Dams across the Brandywine—Wild Animals, Bears, Wolves, Wild Cats—Water Fowl, Disappearance of the Brant.

When the white man first discovered the territory now known as the State of Delaware, it was covered almost entirely with a large growth of forest trees, many of them more than one hundred feet high. Indian corn, various kinds of fruits, and vines, especially the grapevine, were found growing luxuriantly and without cultivation. The latter grew so thick on the site of the town of New Castle, that it was first named Grapevine Point. The country was much better watered than at present, for the clearing off of the woods and draining of the swamps has caused many streams, some of which were navigable, to disappear entirely, and others that once floated vessels that bore the original settlers across the ocean, to so diminish in depth that they will now hardly float the smallest kind of boat without grounding. The harbor or lake back of *Fort Christina*, at the foot of Seventh street, (within the limits of the City of Wilmington,) where the Key of Kalmer lay that brought the first Swedish settlers over, together with the creek that connected it with the Christiana, was more than forty-nine years ago filled up, and workshops are now situated on its site. The Christiana was then over three hundred and fifty feet wider above Wilmington than at present. All the ground, from the foot of Seventh street to the Delaware, now known as Cherry Island Marsh, was under water at high tide, save a small island in the middle, which was covered with cherry trees, from which the marsh derives its name. The marsh opposite the city, on the southern side of the Christiana, was also overflowed at high tide, and the rocks, where the late John K. Kirkman's ship-yard is now, and the Old Ferry Point opposite the foot of Third street, (close to the new Third street Bridge, where the Townsend Iron Works are now built,) were then denominated the Capes of the Christiana, and so inscribed on the records of the county. Between those two points and the Delaware, at high tide, was nothing but a waste of water, save the small spot named Cherry Island. A large stream that eighty-five years ago turned a wheel for sawing marble at the westerly corner of Second and Orange streets, flowed into the Christiana at the foot of Shipley street, and vessels ascended it, and boys bathed in it for some distance above Front street. This stream has now disappeared. The bowsprits of large vessels eighty-five years ago extended over Water street, and the Liberty, a ship of three hundred and sixty tons, was built at the southwest corner of Market and Front streets.[1] Clements' Creek, the little stream that crosses the Newport Pike about a hundred yards from Front street, (or as laid down in the map of the city, at the junction of Justison and Sycamore streets,) that now would not float a batteau, was formerly ascended by vessels to a wharf near the turnpike to take in wood for the Philadelphia market. Vessels ascended the run that flowed down Poplar street, since culverted over, and the great freight house of the Philadelphia, Wilmington and Baltimore Railroad Company, at the foot of Poplar street, is built on the site of Mulberry Dock, where vessels loaded and unloaded their cargoes not more than forty-five years ago. Shellpot, (or Skillpot, as it was formerly called,)

[1] Ferris' Original Settlements on the Delaware. This was a useful and valuable work, written by Benjamin Ferris, an old and esteemed citizen of Wilmington, a member of the Society of Friends. It was published by Wilson & Heald, booksellers of Wilmington, in 1846. It gave a minute historical account of the first settlements on the Delaware, and graphically described the manners and customs of its earlier inhabitants. It also contained an excellent history of Wilmington. Benjamin Ferris died on the 9th of November, 1867, at the good old age of 89 years and 2 months, respected and esteemed by all who knew him. This was the first work ever published devoted mainly to the historical affairs of this State. In 1838 Mr. Huffington published the Delaware Register at Dover, but it was as much of an agricultural and literary as an historical work.

which flows into the Brandywine about three-quarters of a mile from the built up portions of Wilmington, that now would hardly float a longboat, was stated by William Penn to be large enough to contain the whole navy of England.

A similar diminution of the navigable rivers and streams has taken place over the whole State. Every creek between the Christiana and Cape Henlopen has had its navigation injured, and has become much shoaler within the memory of men still living. A bar has been formed at the mouth of Blackbird Creek, in New Castle county. A bar has also been formed at the mouth of Mispillion Creek, in Kent county, which extends near two miles out in the bay, which injures greatly its navigation, and both creeks are much diminished in depth. Stone Wharf Creek, up which the British brought the stone from England (in vessels which sailed from there) to build what is known as the Big Light House on the Atlantic Coast, about a mile from Cape Henlopen, and the brick to erect the dwelling for the keeper, is now, by the action of nature, filled up, and wagons and carts drive over what was once its bed. The site of Synapuxent Inlet, in Sussex county, near Lewes, that used to be navigable for large vessels, can now be drove over with a horse and wagon. In this inlet a French war ship, during the Revolution, took refuge, and landed a quantity of money and arms for the use of our soldiers, as the British then had control of the bay, which were taken up the State on wagons, escorted by a body of Delaware troops under the command of Colonel David Hall. Now not a vestage of it is to be seen. Lewes Creek has also diminished greatly in depth. Thirty years ago large coasters used to winter in this creek, where is now hard and fast land. The place where a large British ship lay, that was captured and brought up this creek during the Revolution, is now a mowing marsh. Lewes formerly had a great deal of trade with New York. Large quantities of grain was shipped from there, and the Creek at certain seasons was filled with vessels waiting to convey it to that city. This trade has been lost from the creek becoming too shallow. Most of the people resident in the neighborhood were formerly engaged in navigation, but the rivers, creeks and inlets diminishing in depth so that they would not float vessels of sufficient size to navigate the ocean, has changed the occupation of the citizens, and they are now cultivators of the ground. Lewes Creek more than one hundred years ago had begun to diminish in depth. We copy the following account of it from Smith's History of New Jersey, a book published in 1765. It was then called the Hoerenkill.[1] The account of this shallowing of the creek, it will be perceived, is derived from a manuscript in the British Museum. It is also given the credit of being the best harbor in the Delaware Bay.

"Two leagues (says the manuscript in the British Museum) from Cape Cornelius, on the west side of the river, near its mouth, there is a certain creek called the Hoerenkill, which may well pass for a

[1] This creek was afterwards called by the English Whorekill.

middling or small river, for it is navigable a great way upward, and its road is a fine road for ships of all burthens, there being none like it for safety and convenience in all the bay, the right channel for sailing up the bay passing it.

"A certain person who, for several years together had been a soldier in the Fort, informed us about the month of June, 1662, being then but lately come from thence, concerning the Hoerenkill, or Harlot's Creek, that along the seashore it was not above two leagues from the cape, and that near the fort, which is at the mouth of it, it is about two hundred paces broad, and navigable and very deep to about half a league upwards, the pilots say generally about six feet of water in going in, but the canoes can go about two leagues higher. There were two small islands in it, the first very small, the last about half a league in circumference, both overgrown with fine grass, especially the latter, and are about half a league distance asunder, and the latter about a league from the channel's mouth. The two islands are surrounded with muddy ground, in which there grows the best sort of oysters, which said ground begins near the first island, for the mouth of the channel has a sandy bottom, being also very deep, and therefore there are no oysters there. Near the smaller island, and higher up, it is as broad again as at the mouth. Near the said fort the channel for a good way runs at equal distances from the sea, having the breadth of about two hundred paces of high downy land lying between them. Near the fort there is a glorious spring of fresh water. A small rill, rising in the southeast part of the country, and falling from a rising hill, runs through this downy land into the mouth of the Hoerenkill, or Harlot's Creek, is for its goodness and fertility named for the very best of New Netherland."[1]

Smith, speaking of the above manuscript, says:

"Soon after English possession it got the name of Lewistown, by which it is mostly called. It is situate at the mouth of Delaware Bay, and is a general resort for pilots waiting to convey vessels up the river. Where the creek is described as deep and sandy is now a mowing marsh. The channel, also, by the Whorekill, then used for vessels to pass, is diminished to about a hundred yards breadth at the mouth. The two islands, one very small, and the other half a league in circumference, are now, the first, supposed to be ten, and the last thirty, times as large as there described, and this alteration in about a hundred years."

Such is the description of Lewes Creek, variously over one hundred, and two hundred, years ago. The islands still exist. They are now several hundred acres in extent. One is now called Green Island.

But though the land has gained on the navigable streams by the narrowing and shallowing of their channels, and the filling of them up altogether, so that vegetation now grows where once large vessels floated, the river, bay and ocean, that bounds the

[1] This was the name of Delaware at that time. The Dutch were then the inhabitants, and our State was part of New York, which was known by the same cognomen.

State, has washed away a large portion of its coast, and the State is now many square miles smaller than when the white man first landed on its shores. It is probable that it is in some places half a mile, and in others two miles, narrower from these encroachments than when first settled. From ten to fifty feet of the State is washed away every year—the fast land becoming marsh, the marsh sand, and the sand becoming covered with water. This is owing to the Northeasterly storms. In the neighborhood of New Castle old residents can point to the foundations of houses, nearly a quarter of a mile out in the Delaware, and in the limits of the same town, after every storm, from the enroachments of the river, the bones and skulls of those buried in a graveyard bordering on it are exposed to view scattered along the beach, and swallowed up by its waters. The lighthouse at the mouth of Mahon's River, in Kent county has within forty years had to be rebuilt three times from the encroachments of the Delaware. The foundation of that first erected is now nearly a quarter of a mile out in the bay. More than half a mile of the point of land that now forms the harbor of Port Mahon is washed away within the memory of the older residents. It is also within the memory of people now living when Duck Creek, and Old Duck Creek, had but one mouth, both flowing into Dona River, which had but one channel at the south end of Little Bombay Hook Island. Bombay Hook was then a part of New Castle county, connected with it by fast land, instead of an island, as now. Duck Creek now flows in a channel at the north of Great Bombay Hook and the main channel of Old Duck Creek, and Little Creek is to the north instead of the south of Little Bombay Hook Island. The old channel at the south of Little Bombay Hook is fast filling up. Both these new channels, viz.: the channel to the north of Great Bombay Hook, and the channel to the north of Little Bombay Hook Island, were made by men for the convenience of getting their boats into the bay. These channels were then washed out and deepened by the force of the current, so that the commerce of the towns of Smyrna is carried through the former, and of Leipsic through the latter. The channel to the north of Great Bombay Hook is known by the name of the Thoroughfare, and is now part of the boundary between New Castle and Kent counties. The old channel is now so filled up that it cannot be used by the vessels employed in the commerce of those towns, and is now only navigated by boats.

On the coast of Sussex county, especially where it it bounded by the Atlantic Ocean, still greater encroachments have been made upon our State, but in some instances, as will be related hereafter, the State has encroached upon the bay.

High hills of sand, from forty to fifty feet high, used to front the bay and ocean, from Grove Bank, near Lewes, in front of Lewes, and extending along the Atlantic Coast to the boundary of the State into Maryland. They extended probably half a mile inland, and were covered with grass and pines. The great storm that about forty years ago devastated our coast, swept down all these hills on the ocean side, made them level, and washed away a great portion of the ground. From this storm the Atlantic encroached on our State about half a mile, and made what was before fast land part of the ocean. During this storm several hundred acres of the farms of Wm. Newbold and John Rhodes were washed away, and many people, and a great number of cattle were drowned. The same storm so encroached on the shores of Rehoboth and Indian River Bay as to change the nature of their waters, and to make them so salt as to destroy the oysters that before abounded there.

The shoals called the Hen and Chickens, situated in the ocean, near Cape Henlopen, now miles out in the sea, it is supposed, was once connected by the fast land with our State. The oldest citizens of the nighborhood recollect it as an island covered with trees. The waters of the Atlantic now roll over it. At extreme low tides the stumps of trees may be seen miles out in the ocean, showing how it has encroached on Delaware, and washed away its shores. At the lowest calculation two miles of the State, where it fronts the Atlantic, and one mile from New Castle to the Cape Henlopen, which was once fast land, is now covered by water. The contrary of this is, however, the case in that portion of the State opposite the Breakwater. This bank of stone, protecting it from the storm, the beach has encroached on the bay, and is now about half a mile nearer the breakwater than it was before that great work was erected. So much so, that the pilots avoid taking vessels out of the channel between the breakwater and Cape Henlopen, and it is thought that in time the beach will so project as to connect itself with the breakwater. Cape Henlopen is also extending itself into the ocean. It now reaches out more than a mile further than it did a hundred years ago. The Big Light, built by the British, to mark the entrance to the Delaware before the Revolution, is laid down in the old charts as being one cable length, or 12 fathoms, (730 feet,) from this point of the cape. Several vessels sailing by this chart, and endeavoring to enter the bay by that distance from the light, were wrecked on the cape. A new light has now been erected on the extreme end of the cape. The distance between the new and old lighthouses is about a mile and a half, so that by this extension of the cape over a mile of land has been won from the ocean. This is a small gain, however. in comparison with what it has robbed us of. But even should its stormy waves in the lapse of eventful time totally destroy our State; wash away both hill and plain, and leave not a vestige of its territory, save what was covered by its waters, Delaware would still live in history and the minds of men; from the glorious deeds of her sons in the Revolution, and from her being the first to adopt the constitution of what *will be* the greatest and mightiest nation the world has ever seen, which now known as the United States of America, may hereafter be the United States of the world. The mortal body of our State may be destroyed, but its soul will live, "till time is old, and hath forgot itself."

Baltimore hundred, in Sussex county, now so populous and well cultivated, was originally a track of worthless land, a mere outlying part of the Cypress Swamp. It has been drained and made valuable. Other great swamps and bogs, such as the Tappahana and Heronton Bogs, have been also drained, thus materially altering the surface of the State.

Every stream, when the Swedes and Dutch first landed here, abounded in shad. They ascended the Brandywine to its head, and were caught in quantities above the City of Wilmington. An act of the Legislature was at one time passed forbidding the erection of dams across the Brandywine, (after the State had been conveyed to Penn,) because such dams would prevent the shad ascending the stream, and thus cause dissatisfaction among the Indians, who, in its season, lived principally upon this fish.

The woods everywhere abounded with deer, bear, wolves, opossums, hares, squirrels, wild turkies, pheasants, wild pigeons, and many other kinds of animals and birds. There was also the wild-cat and the rattlesnake. Wolves were so numerous as to prove a great annoyance, and to cause repeated public efforts to be made for their destruction. Near the margins, and on the surface, of the rivers and creeks, were found beavers, otters, muskrats, terrapin, and other aquatic animals, whilst swans, geese, ducks, brant,[1] cranes, and other water fowl were in great variety and abundance. The shores of the bay were covered

[1] The brant was a water fowl about the size of a duck. There are none now. It totally disappeared about thirty years ago.

with king crabs. So great was the quantity of swans in the bay that the first Dutch settlers named that portion of Sussex county, near Lewes Creek, Swanendale. The cultivation of the ground, and the clearing away of the woods, and draining the marshes, having destroyed the harbor for these animals, they are either entirely extinct, or exist in far lessser numbers. The last wolf has long since gone. Bear and deer, it is still alleged, exist in the Cypress Swamp, although they are rarely seen, whilst there is not a tithe of the wild fowl, fish, or animals there formerly were. They have receded with the Indian before the advancing civilization, and given place to the domestic animals, who are more under the dominion of man.

CHAPTER IV.

The Aboriginal Inhabitants—Leni Lenape or Delawares—They Drive off the original Indians—The Minquas—The Nanticokes—The last Indian leaves Delaware—Bones found at Laurel—Their Government Hereditary—Their Councils—Punishments—Retaliation for Murder by other Tribes—Their Weapons—Mode of War—Cruel Treatment of their Prisoners, they burn them alive and scalp them—Cannibalism—Their Hunting—Their Money—Their Manufactures—Their Boats—Their Dwellings—Their Hospitality—Their Food—Their Marriage—They Practice Polygamy—Their Children—Their Religion—Supposed Tradition of Christ—Their Medicine—Their Heaven—Mode of Burial—Customs thereat—Character of the Indians—Reason why they were conquered by the Whites—They hold a Council to see whether they Massacre the Swedes, decide against it—The only recorded Council of the Aborigines of this State.

THE aboriginal inhabitants of Delaware, at the time the European settlers came amongst them, were known by the name of *Leni Lenape*, meaning in our language, original people. They were called by the English, Delawares, after Lord Delaware, from whom the state and the bay also derives its name. The tradition is that they and the five nations, both emigrated from beyond the Mississippi, and by uniting their forces, drove off and destroyed the primitive residents of the country. Their settlements extended from the Hudson to the Potomac, and their descendants finally became so numerous that near forty tribes honored them with the title of grandfather.[1] These

[1] Thatcher's Indian Biography.

three tribes were in process of time subdivided into many others, as location and convenience required. There were twelve Indian tribes resident within the limits of this State around New Castle. The two most prominent that ruled in Delaware were the Minquas and the Nanticokes. There were probably many others, but history does not record their names. The Minquas inhabited the banks of the Christiana and Brandywine. The Nanticokes the lower end of the State, and the Eastern Shore of Maryland. The last of the Nanticokes took their departure from the Delaware in 1748. (The last Minqua had left long before.) They were then residents of the neighborhood of Laurel. They dug up the bones of their principal chiefs and carried them with them. The bones of the rest of their dead they re-interred in the same vicinity, where they were found by digging about fifty years ago.[1] Each tribe of Indians had a

[1] Huffington's Delaware Register. This Register was a monthly magazine, published by William Huffington, at Dover, in 1838, and printed by Samuel Kimmey. It was one of the best publications of the day, and one of its objects, as stated by the editor in his address in the first number, was, "in a series of numbers, to collect and combine in a sensible form all that can be rescued from the dust of oblivion, from which, at a future day, a history of our State may be written." It is to be regretted that it did not succeed in a pecuniary point of view, for it contained matter of much value, both as regards the history and agriculture of the State. Its publication ceased at the end of the year, and but few copies of it exist. It is, however, found bound in two volumes in the libraries of some of our citizens. The publication of this magazine was the first attempt to preserve and make known our history. Mr. Huffington was a lawyer, a man of refined mind and a good writer. He, at different times, held several important offices. He had been Clerk of the House of Representatives, a member of the

ruler, whom they called a sachem. The office was hereditary, but on the mother's side, in order that no illegitimate children could be placed at the head of the nation. "When a king died, it was not his children who succeeded him, but his brother by the same mother, or his sisters, or her daughter's male children, for no female could succeed to the government." Each king had his council, and nothing of importance was undertaken, such as war, peace, and the sale of land, without being first discussed in council, to which not only were the counsellors called, but also the common people. When they made any treaty of peace or friendship, they gave to those with whom they made it a pipe to smoke, which finally sealed the agreement, as they believed if any one should break it, they would be afterwards visited by some great misfortune. Their punishments generally consisted of fines. If a man committed murder, "he may be forgiven on giving a feast or something else of the same kind; but if a woman be killed, the penalty is doubled, because a woman can bring forth children, and a man cannot." Murder was very uncommon among them until "the white man came, when, under the influence of intoxication from the liquor they sold them, several were committed by the Indians. When they

Legislature, and Mayor of the City of Wilmington in 1848 and 1856. He died at Wilmington in 1860. Samuel Kimmey, the printer, was a man who took great pride in his art, and the typography of the Register was, perhaps, equal to any publication in the country in its day. He was also the printer of the "Revised Code" in 1852. There are few specimens of the typographical art of the time superior to this. He died at Dover, October, 1854, in the fifty-second year of his age.

committed murder under those circumstances, they excused themselves by saying it was the liquor that did it." "When any one of them was condemned to death, which seldom happened, the king himself would go out after him," and as they had no prison to confine the criminal, he generally "fled to the woods." When they find him, "the king first shoots at him, and afterwards those who accompanied him in like manner shoot at him until he was dead." If an Indian kill another Indian of a different tribe, those of the tribe to which the murdered man belonged would send one of their men to kill one of the other tribe, and thus "wars were kindled between them. Otherwise there was no law amongst them, though they generally exercised the law of retaliation."[1]

Their weapons were stone hatchets and bows and arrows, in quivers made of rushes. Their bows were made of the limb of a tree of above a man's length, and their bow strings of the sinews of animals. Their arrows were made of reeds a yard and a half long. At one end was fixed a piece of hard wood about a quarter length, at the end of which they made a hole to fix in the head of the arrow, which was made of black flint stone, or of hard bone or horn, or the teeth of large fishes or animals, which they fastened in with fish glue in such a manner that the water could not penetrate. At the other end of the arrows they put feathers. The flint and stone arrow heads and stone hatchets are still often plowed up in our fields, and

[1] Campanius' description of Indians.

are all that remain to remind us that another race than our own (now extinct) were once lords of our soil, save a few of our streams, that yet retain their Indian names, such as Naaman's, Appoquinimink, and Nanticoke Creeks.

When they went to war, each provided himself with a bow and sufficient quantity of arrows, and placed on his head a red turkey's feather, as a sign they were going to shed blood. After they had carried their wives and children to an island, or other place of safety, they proceeded on their way in a certain order, and when they met their enemies attacked them with great outcries. They thought they had made a great battle when ten or twelve men remained dead on the field. Those who had gained the victory took off the scalps of the enemies they had killed, and carried them away as a warlike trophy. They were very cruel in the treatment of their prisoners. They would cut and slash them alive; cut off their ears, their noses, their tongues and their lips, also their fingers and toes. They also cut off flesh from different parts of the body, and then strewed ashes over the wounds to prevent the blood from flowing, and that their victims might not die too soon. Such an example occurred in the year 1646, when Campanius, the chaplain to Printz, one of the Swedish governors of this State, was in the country. The Indians resident of the Christiana and Delaware had taken one of the Mingoes, and bound him to a tree. They made a large fire round him, and when he was half roasted let him loose. Giving him a fire-brand in each hand, and taking one in each hand themselves, they challenged him to fight, and when at last he could no longer stand, but fell down, one of them sprang upon him, and with his nails cut the skin of his forehead open, and tore off his scalp. The Mingoes and Minquas were often at war. They were also in the habit of eating the flesh of their enemies after boiling it. Campanius also relates that some Indians once invited a Swede to go with them to their habitation in the woods. When he arrived there they treated him to the best the house afforded, and pressed him to eat, which he did. There was broiled and boiled, and even hashed meat, of all which the Swede ate with them. But it did not agree with his stomach, for he threw it up immediately afterwards. The Indians did not let him know what he had been eating, but it was told him afterwards by some other Indians that he had been fed on the flesh of an Indian of a neighboring tribe with whom they were at war, and that that was the broiled, boiled and hashed meat with which he had been treated.[1]

They had a system of fortification, which was by surrounding their villages with palisades. The Minques or Mincks (probably the Minquas) had a fort on a very high mountain, very difficult to climb, about twelve miles from New Sweden,[2] which was the name this State was first known by. This mountain was, probably, Iron or Chestnut Hills, near Newark. The usual employment of the Indians, were fishing

[1] Campanius, 122. [2] Ibid. 127.

and hunting, and shooting with the bow and arrow. Lindstrom, the engineer, who came over with Governor Printz, and who improved the fortifications at Christina (now Wilmington), and at Fort Cassimir (now New Castle), in a manuscript treatise written by him, says:

"As soon as the winter is over they commence their hunting expeditions, which they do in the most ingenious manner. They choose the time when the grass is high, and dry as hay. The Sachem collects the people together, and places them in a circumference of one or two miles, according to their numbers; they then root out all the grass around that circumference, to the breadth of about four yards, so that the fire cannot run back upon them; when that is done, they set the grass on fire, which of course extends all round, until it reaches the centre of the circumference. They then set up great outcries, and the animals fly toward the centre, and when they are collected within a small circle, the Indians shoot at them with guns and bows, and kill as many as they please, by which means they get plenty of venison. When the grass has ceased to grow, they go out into the woods and shoot the animals which they find there, in which they have not much trouble, for their sense of smelling is so acute that they can smell them like hounds. Their Sachem causes a turkey to be hung up in the air, of which the bowels being taken out and the belly filled with money, he who shoots the bird down gets the money that is within it."[1]

[1] Campanius, 128.

The money of the Indians was called wampum. It consisted of beads neatly cut out of white or brown cockle, muscle or oyster shells, through which they bore a hole and string them together on a thread like pearls. Each fathom of wampum was worth five Dutch guilders, reckoning four beads for every stiver. The brown beads were more valued than the others, and brought a higher price. A white bead was of the value of a piece of copper money, a brown one was worth a piece of silver.

Speaking of the money of the Indians, the engineer Lindstrom, says: "Their money is made of shells, white, black, and red; worked into beads, and neatly turned and smoothed. One person, however, cannot make more in a day than six or eight stivers. When these beads are worn out so that they cannot be strung neatly and evenly on the thread, they no longer consider them as good. Their way of trying them is to rub the whole thread full on their noses. If they find it slides smooth and even, like glass beads, then they are considered good; otherwise, they break and throw them away. Their manner of measuring their strings, is by the length of their thumbs, from the end of the nail to the first joint makes six beads, of which the white ones are worth a stiver, or piece of copper money, but the black, or blue ones, are worth two stivers or a piece of silver."[1]

When the Europeans first came, the Indians had no instruments or tools made of iron, but still there was

[1] Campanius, 132.

mechanical talent among them. They could tan and prepare the skins of animals, which they afterward painted. They adorned skins and bed covers with various colored feathers, binding them with a kind of net-work, which, says Campanius, "was very handsome, and fastened the featners very well." They also made light, and warm clothing and covering for themselves of the same material. With the leaves of Indian corn and reeds, they made purses, mats, baskets, and everything else they wanted. They also "made very handsome and strong mats of fine roots, which they painted with all kinds of figures. They hung their walls with these mats, and made excellent bed-clothes of them."

The women spun thread and yarn out of nettles, hemp, and some plants unknown to the white men. Campanius said, "Governor Printz had a complete suit of clothes, with coat, breeches, and belt, made by these barbarians, with their wampum, which curiously wrought with the figures of all kinds of animals, and cost some thousand pieces of gold." Their tobacco pipes were made out of reeds, a man's length. The bowl was made of horn, but sometimes of clay.

Their boats were made of the bark of cedar and birch trees, bound together and lashed very strongly. They also made boats out of cedar trees, which they burnt inside, and scraped off the coals with sharp stones, bones, or muscle shells.

The dwellings of the Indians were made of the branches of trees, twisted together with bark, covered with mats, made of the leaves of the Indian corn matted together. They first put a pole in the centre, and then spread their mats and branches around it, and then cover it above with a roof made of bark, leaving a hole in the top for the smoke to pass through. In the pole they fixed hooks to hang their kettles on, and a large stone to protect themselves from fire, and around it they spread the mats and skins on which they slept. Their principal furniture was the kettle in which they cooked their food. They had no seats, but sat on the ground. The Indians not being accustomed to it, could not sit on chairs. Accordingly, when they visited white men, the tables were always uncovered at the lower end, so that the Indian at meals could get on the table, and sit and eat what was set before him cross-legged. When a white man visited an Indian, he spread his best mats on the ground, and laid before him Indian bread, deer, elk, or bear's meat; fresh fish, and bear's fat, instead of butter. These attentions the Indians expected to be received with thankfulness, otherwise, their friendship would be turned to hatred.

The food of the Indians, was the kind of wild animals which abounded then in the State, which they shot with their bows and arrows. Also fish, which they shot with the same weapons. When the waters were high, the fish run up the creeks and returned at ebb-tide; so that the Indians could easily shoot them at low water, and drag them ashore. They also made bread out of Indian corn, which they crushed between two stones and a large piece of wood. They then moistened it with water, and made it into small cakes, which they wrapped up in corn leaves, and baked in the ashes. When they were traveling or laying in wait for their enemies, they took with them a kind of bread made of Indian corn and tobacco juice, which, says Campanius, "is a very good thing to ally hunger and quench thirst, in case they have nothing else at hand."

Polygamy was practiced amongst the Indians, and marriage was early. As soon as an Indian was seventeen or eighteen years of age, he took one, two, or three wives, according to his ability to maintain them. The woman was expected to be in constant attendance upon her husband. Should she be guilty of infidelity or otherwise misbehave, her husband would at once turn her out with blows, and take another wife in her place. They considered it disgraceful to get married until they had, by some exploit in hunting or war, given proof of their manliness. The girls remained with their mothers, and assisted them in the care of the household, such as making mats and carrying small bundles. When they wanted to get married, which generally happened when they were thirteen or fourteen years of age, they were accustomed to cover their breasts, and wear something upon their heads, by which it was understood that they were ready for a husband. When a warrior or sachem married, his wife wore her clothes for a year, completely covered with strings of wampum, in various figures, with with which her hair, her ears, her arms, and her waist, even down to her knees, were decorated. Her hair was greased and her face painted with all sorts of colors, which gave her a shocking appearance. At the same time the husband's person was similarly adorned. No care was taken of the women when bearing children. She merely laid down behind a bush or tree, and immediately after the birth, both the mother and child would bathe in the water, and the day after be as fresh and well as before. They would then fasten the child with a deer skin to a board, a little longer than itself. From this board it was not detached for many months—the mother always carrying it and suckling it attached in that manner, until it was freed from it, to learn it to walk, which it generally did at nine months old, at which time they gave it a name taken from anything that they thought best suited to it.

The Indian religion acknowledged a supreme being, who made both the heavens and the earth. It also acknowledged an evil spirit, a manetto, manitto or devil. They, however, in contradistinction to the white man, worshipped the evil spirit. Their reason was as follows:

"The Great Sachem in heaven," they said, "is not bad. He does us neither good nor harm, therefore we cannot worship him." "The evil spirit," they said, "is bad, and if we don't do something to please him, he will hurt or kill us, therefore we must worship him." They accordingly offered sacrifices to the evil spirit, in woods. They would erect an altar, and offer upon it meat, fish and tobacco, and all sorts of fruit. This they do whenever they prepare to go into or return from a war. In performing their sacrifices they uttered lamentable cries, with strange contortions of their bodies. One portion of their religion was dancing in circles, with songs and joyful cries. Two of them stood in the middle running to and fro, holding in their hands a hollow reed, or dried skin. The Indians gave to Lindstrom the following account of a portion of their religion, from which *he* appeared to think they had some notion of "Christ and his apostles." They received it by tradition from their ancestors.

"Once upon a time," (Lindstrom said) they informed him, "one of your women came among us, and she became pregnant in consequence of drinking out of a creek. An Indian had connection with her, and he also became pregnant, and brought forth a son, who, when he came to a certain size, was so sensible and clever that there never was one that could be compared to him, so much and so well he spoke, which excited great wonder; he also performed many miracles. When he was quite grown up, he left us, and went up into heaven, and promised to come again, but he never returned. Afterwards there came a big mouth, (meaning an eloquent man), with a large beard, like your big mouths (preachers). There was also another big mouth among us, in former times, but he also went off (pointing to heaven); he promised to come back, but never returned."

When the white men first came amongst the Indians in this State, they were not in the habit of committing excesses in eating or drinking. They lived upon the animal and vegetable productions of the country, and drank nothing but pure water. Therefore they generally lived to an advanced age; many of them to over a hundred years. What sickness they had was trifling, and having a knowledge of the curative properties of many herbs, what diseases occurred amongst them readily yielded to their treatment. Campanius says: "They have a cure for the bite of the large poisonous snakes with which their country abounds, which is truly wonderful. It is a kind of root which they call *snake root;* they chew it and mix it with their spittle when fasting, and lay it upon the wound. It almost immediately reduces the swelling, and soon effects a complete cure."[1]

When an Indian died, his relations and friends brought precious and valuable articles to his grave, in order that he might be provided with everything that he might want when he arrived at the Indian heaven, which they believed laid far to the west, where people went after their death. A country, they said, which abounded in game, and fish, and with everything that might be wished for.

They made their graves round, and lined them with logs, and for their great men with planks and boards. The corpse was placed in it in a sitting posture, and by it was laid its shield and the weapons that belonged to it in life. They tied its hands together, one on each side of its head, and then laid planks or boards underneath it to support it; then filling the grave with earth, they push planks or logs upon it to keep it from the wild animals, and fixed in

[1] See Campanius, 142.

the middle a long painted pole in remembrance of the deceased, on the top of which, if he was a good hunter, they put the figure in wood of some wild animal; if he was a fisherman, that of a fish. For three months afterwards the relations and friends would daily visit the grave, and ask him, with cries and lamentations, why he left them so soon, and why he could not stay longer amongst them, and whether he had not good meat, good drink, and everything else he could wish. They then kept their faces blackened for a year. They were very attentive of their graves, that they might not fall in or be overgrown with grass or bushes, lest the memory of the dead should be forgotten.

Such were the habit, custom, and character of the Indians who inhabited this State, of whom it is believed not one of their descendants now remain alive. They were orators, counsellors, and warriors. Equal in morals and general intelligence to the whites, and, we believe, considering the circumstances, in truth, honor, and honesty their superiors. But they were ignorant of letters; they violated God's law, of not being willing as a race to "earn their bread by the sweat of their face." They were hunters, not agriculturalists, and as all wealth, all science, all knowledge depends upon labor, and those nations or races who employ in that labor their brightest and most acute intellect are the most successful and the most powerful, so notwithstanding the many natural high qualities of the Indian, because his labor was done by his women, who were from the nature of things, weak and ignorant, he was defeated, driven back, and exterminated by the white man, whose labor was performed by the strength of body and intellect of the males of his race, instead of the necessary weakened physical and mental organization of the female. The letters and books from which the white man derived his instruction, the ships which brought him over, the iron of which his cannon and muskets and swords and bayonets were made, and the forts which protected him, and the implements with which he tilled the earth, producing from a small space a large crop, were all the results of male labor. It was the possession of these things that enabled him to conquer his red brother. Without them he would have been as helpless as the Indian. Without labor he could not have had them. Therefore, the real cause of the fall of the Indian, in his conflict with the white race, was his contempt for labor, and placing it upon his women. As the white man cleared the woods and plowed the fields, the game having no cover, retreated from his advancing footsteps. The Indian, depending mostly on game, went back with the animals, which the white man drove to the receding wilderness. So that even had there been no war between the races, the Indians must have been driven to the wilderness as the white man advanced, which every day was, and is now, receding to the westward.

These accounts of the Indian we have got mainly from Campanius, as he is the only writer that has dwelt at any length upon those of that race who inhabited the territory now comprised in the boundary of Delaware. There was probably but little difference in character between them and the other Indians that inhabited this Continent; but as his description apply especially to the Indian inhabitants of this State, and those residing in our immediate vicinity in Pennsylvania and New Jersey, we prefer his description to any other.

Although the following council of Indians might have been more appropriately introduced under the events of 1645, yet we relate them here, as illustrating the character of the Indians. As regards the place where the council was held history does not inform us. Printz was governor, of what is now our State, at the time. It was called by the head sachem *Matta Horn*,[1] to know whether the then inhabitants of Delaware, principally Swedes (though there were some Dutch) should be destroyed. The sachem calls his son, Agga Horn, and a dialogue occurs between them, as follows:

Father Matta Horn.—Where are the Swedes and the Dutch?

Son Agga Horn.—Some of them are at Fort Christina, and some at New Gottenberg.

Father.—What do the Swedes and the Dutch say now?

Son.—They say, why are the Indians so angry with us? Why do they say they will kill all of us Swedes,

[1] This Sachem owned the territory on which the city of Wilmington is built. The grounds on which Fort Christina was built was purchased from him, and on that ground was his wigwam. He is sometimes called Matta Hoon.

and root us out of the country? The Swedes are very good. They come in large fast sailing ships, with all sorts of fine things from Swede's country, or old Sweden.

F.—Go round to the other chiefs and to the common men, and hear what they say.

S.—They say, you Indians and we (Swedes and Dutch and English) are in friendship with each other. We are good men. Come to us. We have a great deal of cloth, kettles, gunpowder, guns, and all that you may want to buy.

F.—I understand. What do you say about this, Agga Horn, my son?

S.—I say that I think it best not to fall upon them, because the Swedes are skillful warriors.

F.—My son, you must go about here and there, to our good friends, the officers and common men, and engage them to come immediately here to me, that we may consult together as to what we shall do.

S.—It is well, I will go.

F.—Do that, but don't be long away.

The son comes again and salutes his father.

S.—My father, Matta Horn (that is), Good bye, father, Matta Horn.

F.—Yes, here I am my dear son, Agga Horn.

S.—Father Matta Horn, I have done what you ordered me.

F.—Well, my son, what answered the officers.

S.—They answered that they would come here to us, the day after to-morrow.

F.—You, my son Agga Horn, may go with the men to shoot some deer in the woods. Perhaps the good gentlemen may be hungry when they come.

S.—I understand that well, I will go immediately out hunting.

After being hunting, he returns with venison.

F.—Have you been hunting?

S.—Yes, I have.

F.—What have you done?

S.—We have killed two elks, and as many deer as will be wanted.

F.—Have you shot no turkeys?

S.—I shall have also, twelve turkeys.

F.—Enough, enough.

The people are now assembled in Council.

Sachem.—Are you here, good friends?

Warriors.—Yes, we are.

Sachem.—That is well, you are welcome. Set down and rest.

Warriors.—With pleasure, for we are much tired.

Sachem.—Are you also hungry?

Warriors.—Yes, may'be we are hungry.

Sachem.—I know you have gone a great way, so you must be very hungry. We shall have meat presently.

Warriors.—That will do for us.

Sachem.—Here, you have to eat. Eat all, ye good friends.

Warriors.—Yes, we will do our best. Give us meat.

Sachem.—Do you also want drink?

Warriors.—Yes, give us drink. This is sweet and good water. We are now well satisfied. Thanks, thanks.

Sachem's Speech to the Warriors.—My good friends,

all of you don't take it amiss that my son has called you to this place. The Swedes dwell here upon our land, and they have many fortresses and houses for their habitation. But they have no goods to sell to us. We can find nothing in their stores that we want, and we cannot trade with them. The question is, whether we shall go out and kill all the Swedes, and destroy them altogether, or whether we shall suffer them to remain? Therefore, I am glad that you came here, that we may consult together on this subject. You chiefs and warriors, what advice do you give? What shall we do with the Swedes? They have no cloth, red, blue, or brown. They have no kettles, no brass, no lead, no guns, no powder. They have nothing to sell us; but the English and Dutch have got all sorts of merchandize.

Some of the Chiefs answer.—We are for the Swedes, we have nothing against them.

Another Chief answers.—It would be well to kill all the Swedes; for they have nothing in their stores, for which we can trade with them.

The Common Warriors answer.

A common warrior says: Wherefore, should we kill all the Swedes, and root them out of the country? They are in friendship with us. We have no complaint to make of them. Presently they will bring here a large ship full of all sorts of good things.

Others answer.—You talk well, we common warriors agree with you. Then we shall not kill all the Swedes, and root them out of the country.

Others reply.—No, by no means. For the Swedes

are good enough, and they will shortly have here, a large ship full of all sorts of goods.

The King's decision.—Right so. We, native Indians, will love the Swedes, and the Swedes shall be our good friends. We, and the Swedes, and the Dutch, shall always trade with each other. We shall not make war upon them and destroy them. This is fixed and certain. Take care to observe it.

The whole meeting answers.

We all agree it shall be fixed and certain.

Now, we are going home.

Yes, farewell.

Whither are you going?

To our plantations.

I understand.

The maize is now fully ripe.

Yes, it is certainly ripe.

Now then, fare ye well.[1]

Such is the account given by Campanius of the council held by the Indians, to decide whether they should attempt to massacre the first settlers of Delaware. It is the only recorded proceedings in existence, of any council held by the Indians who inhabited this State, or of any meeting of theirs, that had any relation to its inhabitants. This council will be alluded to in our history, hereafter.

[1] See Campanius, 153, 154, 155, 156.

CHAPTER V.

FROM A.D. 1492 TO 1606.

Discovery of America by Columbus—Of the Continent by John Cabot—Sebastian Cabot sails from Labrador to Virginia—Passes the Southern Boundary of Delaware—Makes several other voyages—Made Grand Pilot of England—Verrazani touches the Continent in the latitude of Wilmington—Grant to Sir Humphry Gilbert—He touches the Continent near the Kennebec—No grant of Delaware—Lost on his return to England—Grant to Sir Walter Raleigh—He has the right to Delaware when he discovers it—He does not do it—He assigns the right to merchants in London—James I. claims the land between the 34th and 45th degrees of latitude—Grant to North Virginia, from 41st to 45th degree of latitude—To South Virginia 34th to 38th degree—Delaware not included.

[1492] THE first discovery of the Western hemisphere was made by Christopher Columbus, a Genoese, who was employed by Ferdinand and Isabella (the King and Queen of Spain) to discover a new route to the Indies. He sailed from Palas, in Spain, on the 14th of August, 1492, old style,[1] and landed at St. Salvador, or Cat Island, on the 12th of October following. This discovery of Columbus was the cause of the settlement of this State by its pre-

[1] There are nine days difference between the old and new styles, the new style being nine days later. English and Swedish date up to 1752 are old style, and nine days must be added to them to correspond with our present mode of reckoning. Dutch dates are new style, or dates now in use. They adopted the new style about 1600. The English did not adopt it until 1752.

sent people. As the discoveries by him incited the enterprise of the citizens and navigators of England and Holland, the former of whom in subsequent voyages discovered the continent of which our State is a part, and the latter the river, on the banks of which it is situated. In May, 1497, or about five [1497] years after Columbus saw the first island of the New World (as the Continent of America is called), John Cabot, a Venetian, under the authority of the English King Henry VII., discovered the continent. On the 21st of June he first saw what was supposed at that time to have been the Island of Newfoundland, but what is now thought to have been the coast of Labrador. He soon afterwards returned to England. The following year his son Sebas- [1498] tian Cabot, who was with him on his first voyage, and born in Bristol, in England, and therefore an Englishman, made a second voyage, and explored the continent from Labrador to Virginia, and some say to Florida. He thus sailed past the southern shore of this State, on the Atlantic. After several other voyages he returned to England, during the reign of Edward VI., and as a reward for his services was appointed grand pilot of the kingdom. Several other voyagers made discoveries in America, but it is not our purpose to allude to any but those that have in some manner been connected with the State of Delaware. In 1524, or twenty-seven years [1524] after Sebastian Cabot had sailed past it on the Atlantic, John Verrazani, a Florentine of celebrity, in the employment of the French, discovered

the continent in the latitude of Wilmington, in this State. He must therefore have touched or observed it at what is now called the Long Beach, in New
[1578] Jersey, near the town of Tuckerton. In 1578 Queen Elizabeth of England gave to Sir Humphrey Gilbert an open or patent letter for "*all such remote heathen and* barbarous lands as he should discover in North America, and of which he should take possession; these lands not having been occupied before by any other Christian power." She vested in him and his heirs the right of property, and guarantied that all who should settle there should enjoy the privileges of free citizens and natives of England. He was to acknowledge the sovereignty of England, and pay one-fifth of all the gold obtained. Under this patent he made several voyages during the year 1579 and 1583, and touching at the Island of New Foundland, sailed as far south as the Kennebeck, but it does not appear that he had himself any grant, or was in any way connected with Delaware, though some geographers place down the whole territory between Florida and New Brunswick as being the "*Remote and Heathen Lands*" *patented by Queen Elizabeth* to Sir Humphry Gilbert in 1578. This map is so laid down in Willard's History of the United States. This grant would, of course, include Delaware, but as the patent granted only included such lands as he discovered, and he did not sail further south than the mouth of the Kennebeck River, in Maine, he could have never had any jurisdiction over our territory. However, upon the death of Gilbert, (a noble, gallant

and Christian sailor, who was lost at sea in a little vessel of ten tons, called the Squirrel, on his return to England,) a patent was granted by the same queen to the celebrated Sir Walter Raleigh, his brother-in-law, for all the land he should discover between the 33d and 40th degree of north latitude. This gave him the right to all the territory, *when* he found it, from a short distance south of the Santee River, in South Carolina, to a point a few miles north of Tom's River, in Ocean County, New Jersey, and also a mile or so north of Philadelphia, at a point between Philadelphia and Germantown. Raleigh sent several vessels to America, which discovered Albemarle and Pamlico Sounds, in North Carolina. He there established a colony on Roanoke Island, which was destroyed. But it does not appear that either he, or any of those under him, ever sailed as *far north as Delaware Bay*, and all his connection with this State was a *right* to discover and possess it; a *right* which he never exercised.

Soon after Raleigh assigned his patent to [1589]
a company of merchants in London. Seven years after this assignment to the London merchants James I. of England, claiming all the land [1606]
between the 34th and 45th degrees of north latitude, (or from Cape Fear River, within a mile or two of the southern boundary of North Carolina, to the St. Croix River, which divides the northern boundary of the United States at the State of Maine from the British Colony of New Brunswick,) divided it into two districts, which he called *North* and *South*

Virginia. North Virginia included from the 41st to the 45th degree, and contained the whole of New England, nearly the whole of the State of New York, a small part of New Jersey, and that portion of Pennsylvania north of a line drawn through it from east to west, from about Stroudsburg, in Monroe County, on its eastern border, to New Castle, in Lawrence County, on its western border. This he granted to the Plymouth Company, composed of "knights, gentlemen and merchants." *South Virginia* included from the 34th to the 38th degree of latitude, and contained the territory between the mouth of the Cape Fear River, near the boundary of North and South Carolina, and the boundary of Virginia and Maryland. This was granted to the London Company, composed of "noblemen, gentlemen and merchants," mostly resident of the City of London. The intermediate district, from the 38th to the 41st degree, comprising the States of Delaware and Maryland, and the largest part of Pennsylvania, nearly the whole of New Jersey, Manhattan Island, on which the present city of New York stands, together with Staten and nearly the whole of Long Island, was open to the settlement of both companies, but neither was to come within one hundred miles of the other. These grants were thus made three years before the discovery of Delaware River, by either the English or any other nation. Under these companies both Virginia and New England were settled. The Plymouth or North Virginia Company, however, fourteen years afterwards, succeeded in getting their charter modified, and their territory extended, by an additional grant of one degree of latitude from the 40th to the 41st degree, thus bringing under their dominion the whole of New York and nearly the whole of Pennsylvania, and near two-thirds of New Jersey.

CHAPTER VI.

FROM 1609 TO 1614.

Employment of Henry Hudson by the Dutch East India Company, to find a Northeast passage to China—Sailed from Texel, in the yacht "Half Moon"—His Discovery of Cape Cod—Supposed Discovery of Chesapeake Bay—Discovery of Delaware Bay—Log of Robert Jewett, Hudson's Mate—Discovery by Capt. Argall—Visit of Lord De-la-war, in the Bay from which it derives its Name—Abandonment of Hudson by his Mariners in Hudson's Bay—Sketch of the Life of Hudson—Recorded Names of Crew of Half Moon—Samuel Purchase, First Writer on Delaware.

[1609] To Henry Hudson, an Englishman, in the employ of the Dutch East India Company, belongs the honor of first discovering the State of Delaware. He certainly never landed, but sailing into the Bay from which the State derives its name, he undoubtedly obtained a sight of our shores. He was engaged by the East India Company to find a Northeast passage to China. That measure at that period, obtaining a large portion of the attention of the scientific and commercial portions of the civilized world. Accordingly, he was engaged by that Company, as captain and supercargo of the ship or yacht "Halvemann," (or Half Moon,) 40 lasts or eighty tons burthen. She left the Texel, April 9, 1609,[1] and sailing toward the Northeast, endeavored to make a passage to China in that direction, but changed his

[1] Broadhead's Address, N. J. Historical Col.

course owing to the ice, and stood over toward what then was called New France, now, the British possessions of North America. He passed the Banks of Newfoundland, in latitude 43° 23′. He made the land in latitude 44° 15′, and went on shore at a place where there were many of the natives with whom, as he understood the French came every year to trade. This place is supposed to be the mouth of the Penobscot, or a small French settlement, now Annapolis.

From thence he took his course to the South, running S. S. W., and S. W. by S., where he again made land, in 41° 43′, which he supposed to be an island, and gave the name of New Holland, but afterward discovered that it was Cape Cod. Pursuing his course toward the South, he again saw land in 37° 15′. The coast was low, running North and South, and opposite to it lay a bank or shoal, within which was a depth of 8, 9, 10, 11, and 6½ fathoms, with a sandy bottom. This he called Dry Cape (supposed to be Chesapeake Bay, and Cape Charles). Changing his course to the northward, he again discovered land in latitude 38° 9′, where there was a white sandy shore, and within appeared a thick grove of trees, full of green foliage. His direction of the coast was N. N. E., and S. S. W., for about 24 miles, then North and South for 21 miles, and afterward S. E. and N. W., for 15 miles. They continued to run along this course to the North, until they reached a point from which the land stretches to W. N. W., where several rivers discharge into an open bay. Land was seen to the E. N. E., which Hudson at first took for an island, but it proved to be the main land, and the second point of this bay, in latitude 38° 54′. This was, without doubt, Cape May, now laid down in latitude 38° 57′, varying only three minutes from the observation of Hudson; the remainder of the description applies well enough to the Delaware Bay and River, now, first discovered by the Dutch. Standing in upon a course N. W. by E., they soon found themselves embayed, and encountered many breakers, and stood out again to the S. S. E. Hudson supposed that a large river discharged into the bay, from the strength of the current that set out, and caused the accumulation of sands and shoals. Convinced that the way to China did not lay in that direction, they then continued along the coast toward Sandy Hook.[1]

[1609] The following is from the log-book of Robert Jewett, the mate, who gives the following account of the discovery of Delaware Bay.

"Friday, Aug. 28. Fair, and hot weather, wind S. S. W. In the morning, at 6 o'clock, we weighed and steered away north 12 leagues until noon, and came to the point of the land; and being hard by the land in five fathoms, on a sudden we came into three fathoms, and we bore up and we had but ten feet water, and joined to the point. Then, as soon as we were over, we had 5, 6, 7, 8, 9, 10, 12, and 13 fathoms. Then we found the land to trend away N. W., with a great bay and river. But the bay we found shoal, and in the offing we found ten fathoms,

[1] De Laet's Description, N. Y. His. Col.

and had sight of beaches and dry sands. Then we were forced to stand back again, so we stood back S. E. by S., three leagues, and at 7 o'clock we anchored in eight fathoms of water, and found a tide set N. W. and N. N. W., and it rises one fathom, and flows S. S. E. And he that will thoroughly explore this great bay, must have a small pinnace, that must draw but four or five feet water, to sound before him. At 5 in the morning we weighed, and steered away to the eastward on many courses, for the more northern land is full of shoals; we were among them, and once we struck, and we went away and steered away to the S. E., so that we had 2, 3, 4, 5, 6, and 7 fathoms, and so, deeper and deeper."[1]

From this it will be seen, that Hudson was the first person that discovered this State, and that the 28th day of August, 1609, was the first time its shores were seen by civilized man. One year later, it is alledged, that Sir Samuel Argall,[2] after- [1610] ward Governor of Virginia, visited the Delaware Bay, and named it Cape Delaware, after Lord De-la-war, the Governor of Virginia. In his report he states, that he caught halibut, cod, and ling fish in the Bay.[3] The year afterwards, Lord De- [1611] la-war himself visited the Bay, on his voyage homeward. It was after this called by the English Delaware Bay. The Indians called it Chickohockie.[4]

[1] N. Y. His. Col.

[2] It was Argall that seized on Pocahontas, as a hostage for the good conduct of her father, the Indian chieftain, Powhattan.

[3] Strachey.

[4] Anderson's History of Colonial Church.

To Hudson, therefore, is due the honor of its first discovery, and although his name is not applied to it, it still lives in that of the great river, on the banks of which is seated the great metropolis of this continent, and the great bay of the North, which has proved both his "tomb and his monument."[1] In other words, Hudson River, on which the great city of New York is situated, and the great Hudson's Bay, in the British possessions of North America, where he was abandoned by his treacherous sailors, and never afterward heard of.

The Delaware Bay (and consequently the State of Delaware), was discovered by Hudson six days before he entered the Hudson River. As we have before said, the 28th of August, he sailed in and explored the waters of the great Bay, from which this State derives its name; whereas the Half Moon did not anchor within Sandy Hook until the evening of the 3d of September. New York, to use the words of an eminent descendant of Delaware, is accordingly, Delaware's younger sister.[2]

Of the birth, parentage, home, boyhood, and early days of the manhood of the discoverer of Delaware; nothing is known prior to the 19th of April, 1607, when he suddenly appears upon the stage of action as a captain in the employ of the Muscovy Company, an English Company, of which another Henry Hudson

[1] Bancroft, 265–275, 19th Edition.

[2] Lecture delivered by Jno. Meredith Read, before the Historical Society of Delaware, October 13, 1864, on the life of Hudson. Published by the Historical Society of Delaware.

(supposed to be his grandfather[1]) was one of the founders, formed for the purpose of trading between England and Russia. The original name of the family was supposed to be Hodgson (derived from Hodge's son—the original bearer of the surname being the son of Hodge), and thus changed to Hodgson, and from thence to Hudson by various modes of spelling. He first commanded a ship called the Hopewell, in which he was sent to discover a route to China by the way of Spitzbergen and the North Pole in April, 1607. In 1608 he made a second voyage for a similar purpose, for the same company, which resulted in making known a portion of Nova Zembla. In 1609, in the service of the Dutch East India Company, he discovered New Netherlands (a part of which was the State of Delaware). His mate desired that he should winter in Newfoundland and search for a northwestern passage. But as his crew were mutinous, and had savagely threatened him, and as many of them were ill and sickly, they returned homeward. On their voyage they put into Dartmouth in England on the 7th of November. Hudson and the other English were here commanded not to leave England but serve their own country.[2] The Half Moon[3] returned to Amsterdam after eight months detention.[4]

In the preceding month of April, Hudson sailed

[1] Lecture by John Meredith Read. [2] Purchase's Pilgrimages.

[3] This vessel, the first that ever entered the Delaware, was wrecked at the Island of Mauritus in 1615. Broadhead i., 43.

[4] Stowe's Chronicle, 509, 510.

under English auspices again to search for a north-west passage. He wintered in latitude 52, and sailed up to latitude 60 along the western shore of the Hudson Bay. Here his crew mutinied. They had been absent from home ten months, with provisions for only eight, and during their whole voyage they had met with but a single man, an Indian armed with a cris or poinard. He brought them an animal which they ate, but having badly treated him, he went away and never returned. Now, although, "he had divided even with tears his last bread with his men, yet on a midsummer's day in 1611, his ungrateful crew thrust him into a frail boat with his son,"[1] John Hudson, and left them to their fate. The crew then returned by the way they had come, and reached their home in September, 1611, where they were thrown into prison. Three ships were fitted out and sent in search of Hudson by the King, the Prince of Wales, and some merchants, but the unfortunate discoverer of Delaware was never heard of more.

All published accounts[2] of Hudson are derived from "Purchase's Pilgrimages; or, Relations of the World," an unfinished work giving an account of the voyages of the early navigators.[3]

[1] Bancroft.

[2] J. M. Read, Jr.

[3] The Rev. Samuel Purchase was a London clergyman. He is entitled to the honor of being the first author who wrote of our State. He was a philosopher, historian and theologian, widely known for his writings, especially for his large volumes pertaining to the East and West Indies. The publishing of his works brought him in debt, but he died not in prison, as stated, but in his own house.

Besides Hudson, Robert Juet and John Coleman, are the only recorded names of the crew of the Half Moon that visited Delaware Bay. Hudson's manuscripts are lost, and the only written account of his visit to the Delaware, is that of Juet, who lived at Limehouse.

Delaware Bay and River has received different names from the various nations who have at different times inhabited it. By the Indians it was called Pontaxat, Chickohockee, Mariskitten and Moherishkisken, and Lenape Whittuck. The Dutch called it Zuydt or South River, Nassau River, and Prince Hendrick's or Charles River; the Swedes, New Swedeland Stream; the English, Delaware; Heylin, in his Cosmography, calls it Arasapha. It has also been known as Newport and Godyn's Bay.

CHAPTER VII.

FROM 1614 TO 1621.

Passage of General Edict by States of Holland in favor of Discoverers of New Lands—Terms of the Edict—Fitting out of vessels under it—Building of Block's vessel the Fortune—Building of the Yacht Restless, the first vessel built in the United States—Naming of Cape May from Capt. Cornelis Jacobsen Mey—Return of the vessels to Holland—Hendrickson sails up the Delaware to the Schuylkill—Hendrickson the first who Landed in the State of Delaware—He purchases Indians from the Minquas—Block, May and the rest form a Company and Petition the State General for Confirmation of the Privileges promised to Discoverers by their Edict—Their Petition granted—Death of Lord Delaware from whom the River and State derives its Name—His antecedents and Family.

[1614] AFTER the discovery by Hudson in 1609, no steps were taken by Europeans to settle the shores of the Delaware until 1614, at least none known at the present day, as the document relating to events between those periods were destroyed in Holland. In that year in consequence of petitions being presented to the States of Holland by "many merchants interested in the maritime discovery" to what, in the terms of the petition, were called the "High and Mighty States General of Holland," a general edict was passed in favor of all persons who should discover "any new courses, havens, countries, or places, of the exclusive privilege of resorting to and frequenting the same for *four voyages*." If any violated the provisions of the edict, they were to forfeit their vessels and be fined 50,000 Netherland ducats, which were to be given to the discoverer whose rights they had infringed upon. The discoverer in fourteen days after his return was required to deliver to the State "a pertinent report of his discoveries." If one or more companies were to discover the same countries "within the same time, then they were unitedly to enjoy the privilege of the four voyages, the time when they shall cease to be determined by the States, who were also to settle any differences arising."[1]

Under this edict, there were five vessels fitted out by merchants of Amsterdam, viz., the Fortune belonging to Hoorne, by Captain Cornelis Jacobson Mey; the Tiger, commanded by Captain Hendrick Cortienson; the Fox, Captain De With; the Nightingale, Captain Volkersten, and another vessel, named the Fortune, commanded by Captain Adrien Block. These vessels sailed to the mouth of the Manhattan River, where Block's vessel was unfortunately destroyed by fire. To supply the place of his burnt ship, he built at a small island, near the mouth of Long Island Sound, on the coast of Rhode Island (and now named after him Block Island), a yacht of 38 feet keel, 44½ feet long, and 11 feet wide, which he called the "Onrest" or Restless. She was when finished about 16 tons burthen. This was the first vessel built in this country by Europeans. With the exception of the Fortune, Captain Mey, all these vessels sailed to

[1] Hist. Doc., translated by O'Calligban.

the eastward. Captain Mey sailed south and arrived at the Delaware Bay, and it is from him that the eastern cape of the Delaware derives its present name of Cape May. The southern cape was named after his first name Cape Cornelius, but it was afterwards changed to Cape Henlopen, the name it at present bears. Shortly after this all the vessels returned to Holland, with the exception of the yacht Restless, which was placed under the command of Captain Hendrickson. She was left to make a more minute examination of the country, and was the first
[1616] vessel to explore the Delaware Bay and River. In it Hendrickson sailed up the river as high as the Schuylkill. He was consequently the first whom it is known, beyond a doubt, discovered this State and landed on our shores. Hudson had merely a view of our coast, from where our southern boundary touches Worcester County, Maryland, to about where the town of Lewistown now stands. There is no evidence that either Argall or Delaware went on shore. But Hendrickson landed on our soil, and made purchases of some prisoners taken in battle from the Minquas who inhabited the banks of the Christiana. Therefore, to him, we think, belongs the real honor of being the first discoverer of the State of Delaware. In his report, which is among the Holland Documents, and which, however, gives but little information, he speaks of "having discovered and explored certain lands, a bay and three rivers, situated between 38 and 40 degrees, in a small yacht of sixteen tons burthen, named the "Onrest" (Restless),

which had been built at Manhattan. He also states, that "he bought three native inabitants from the Maquas (Minquas) and Mohicans, who held them in slavery, for whom he gave in exchange kettles, beads and merchandise." He also furnished a very curious map (a fac simile of which is now at Albany, N. Y.) drawn on parchment, about two feet long and eighteen inches wide, and "executed in the most elegant style of art," showing "very accurately the coasts from Nova Scotia to the Capes of Virginia. Hendrickson applied to the States of Holland for the privileges promised by the edict passed by them, and on the faith of which, he made his discoveries, but from some cause he was unsuccessful in his application.[1] The bay and rivers, spoken of by Hendrickson as discovered by him, were undoubtedly the Delaware Bay and River, the Christiana and the Schuylkill. The Delaware was the river on which he sailed. The Christiana the one from which he purchased the slave Indians from the Minquas who inhabited its banks, and the Schuylkill, the one that marked the limit of his voyage up the first mentioned river.

After Block, Mey, and their fleet returned to Holland, they formed themselves into a company, and on the 11th of October they petitioned the States General for a special edict in their favor, agreeable to the terms of the general ordinance of the 27th of March. They stated that at great expense and heavy damages to themselves, arising from loss of vessels during the last year, they had, with five ships owned by them,

[1] Broadhead, 18; Hist. Doc., 59; O'Callighan, 18.

discovered and explored certain new lands lying in America between New France and Virginia, in the latitude from 40 to 45 degrees, which they called "New Netherlands." They also presented a map of the newly discovered country, which amongst other things, contained a faithful delineation of the Hudson River as far as Albany, which was made within five years after the discovery of that river by Hudson (a fac simile copy of which is also preserved amongst the records of New York). The State General after hearing the report and examining their map, granted to Captains De With, Block, Volkersten and Mey, the discoverers, now united into one company,[1] with the privilege "exclusively to navigate to the said newly discovered lands lying in America between New France and Virginia, the coast of which is situated in latitude from 40 to 45 (now called New Netherlands), for five voyages, within the period of three years, commencing the first day of January, 1615." None others were allowed the privilege of navigating to or trading with those countries under penalty of the confiscation of the vessels and cargoes, and a fine of 50,000 Netherland ducats for the benefit of the discoverers. This decree was dated at the Hague, October 11, 1814. They thus granted to these navigators, what King James the First had claimed eight years before, and granted the most of it, viz., between 41 and 45 degrees, to the North Virginia Company, in 1606. We have no evidence that the vessels of this company ever traded on the Delaware. Their privileges expired by their own

[1] Broadhead Address. Holland Doc.

limitation in 1618. An application for their renewal was partially granted and for limited periods.[1] The same year the company's privilege expired.

Lord De-la-war from whom the bay and State derives its name, died. Some say, off the Capes of Delaware; others, off the Western Isles. He was on a voyage from Virginia to England. It has been asserted that he was poisoned.[2] This, however, we do not believe. There were three hundred persons on board the vessel with him at the time, sixty of whom also died. [1618]

Lord De-la-war's real name was Sir Thomas West (West being the family name of the De-la-wars). He was the third son of Lord De-la-war, and we suppose, out of courtesy, received the title of his father. In 1602 he married the daughter of Sir Thomas Shirley. The name of Shirley, the ancient seat on James River, Virginia, may be traced to this source. He was the first Governor of Virginia, and one of the best. His name first appears in a commission appointed in the reign of James the First, "for inquiring into the cause of all such persons as should be found openly opposing the doctrines of the Church of England." Persons descended from the West stock are yet to be found in Virginia bearing the name. West Point in that State derives its name from this source. Earl Delaware, who lived in England a few years ago (and probably may be yet living) is a descendant of his. All, however, that he had to do with our State, was the honor of giving us a name.

[1] Holland Document.

[2] Beverly's Virginia.

CHAPTER VIII.

FROM 1621 TO 1629.

Charter of the Dutch West India Company, the United Company of the United Netherlands—Licenses granted to Trade with New Netherlands, its Boundaries, which included Delaware—English hear of Dutch Trading on the Delaware—Vessel sent by them runaway with by her crew—Information of Dutch Trading sent to England—English Ambassador remonstrates with the Dutch Government—Sailing of an Expedition for the South (Delaware) River with Colonists—Building of Fort Nassau—Arrival of Governor Minuit New Amsterdam—William Usselincx presents the plan of a Swedish West India Company to Gustavus Adolphus—Granting of the Charter, its principal features, delight of the Swedes at the enterprise, they eagerly subscribe, their attempt at Settlement stopped by the War—Birth of Queen Christina, in whose reign the State was first settled.

[1621] The privileges of the first company, or original discoverers of Delaware and other portions of this continent, having expired by limitation, and the trade thus becoming free to all, the celebrated West India Company was chartered, under whose auspices the first settlements were made on the banks of the Delaware, and within the limits of this State. The charter provided that for the space of twenty-four years no native inhabitants of the United Netherlands should be permitted to sail to or from the said lands, or to traffic on the Coast of Africa, from the tropic of Cancer to the Cape of Good Hope, nor in the countries of America or the West Indies, beginning at the south end of Terra Nova by the Straits of Magellan, La Maire, or any other straits and passage situate thereabout, to the Straits of Anian, as well on the North Sea, as on the South Sea; nor any islands situate on the one side or the other, or between both, nor on the western or southern countries, reaching, lying, and between both the meridians, from the Cape of Good Hope, in the east, to the east end of New Guinea, in the west, but in the name of the United Company of these United Netherlands, "under penalty of forfeiture of goods and ships found for sale on the above coasts and lands. The charter to operate from the 1st of July."

The company may in the name and authority of the State make alliances, contracts, &c., with the natives of the countries mentioned, build forts, "appoint and discharge governors, equip armies," appoint "officers of justice, and other public officers, &c.;" "they must advance the peopling" of these countries, &c., and transmit a report of such contracts and alliances, and "the situation of the fortresses, &c. taken by them. The States to approve of instructions to governors," and to grant the commissions, with various other regulations of their internal concerns.[1]

The company had five branches, or chambers, in different sections; but the principal was at Amsterdam. The board governing consisted of nineteen members, which was usually denominated the College of nineteen, of which Amsterdam furnished

[1] Hazard's Historical Collection.

eight members, the State General one, and Zealand, Maeze, Friesland, the North Department and Groeningen the remainder.

The West India Company did not commence operations under their charter for some time after its grant. Licenses were, however, granted to several persons to send out two vessels to truck and trade with the natives in newly discovered countries between latitudes 40° and 45°, called "New Netherlands," and to the adjacent territories, together with a great river lying between 38 and 40 degrees of latitude.[1] The *great river was undoubtedly the Delaware*, as it is the only *great river* lying between those two degrees. These vessels were bound to return with their cargoes before the following 1st of July. We have no account of their visit to Delaware. But information from several hands had reached the Virginia Company that the French and Dutch carried on a very profitable trade with the Indians on Delaware and Hudson Rivers, which they supposed "were within their grant, and then esteemed parts of Virginia. The Company therefore this year resolved to vindicate their rights, and not to permit foreigners to run away with so lucrative a branch of their trade. One Captain Jones was accordingly sent upon the voyage, but by the wickedness of him and his mariners, the adventure was lost, and the whole project overthrown," after having been supported by the Earl of Southampton and Sir Edward Sandys, who each subscribed £200.[2]

[1] Holl. Documents.

[2] Stith's History of Virginia.

Confirmation of this attempt of the Dutch to trade with and occupy territories on this continent was immediately sent by the Virginia Company to the English Government, who at once directed their Ambassador at the Hague, Sir Dudley Carleton, "to bring the subject of the Dutch Plantations in North America to the special notice of the State General." The English Privy Council say, "Whereas, His Majesty's subjects have many years since taken possession of the whole precincts, and inhabited some parts north of Virginia, (by us called New England,) of all which countries his Majesty hath in like manner, some years since, by patent granted the quiet and full possession unto particular persons, nevertheless, we understand that the year past, the Hollanders have entered upon some part thereof, and have left a colony, and have given new names to the several ports appertaining to that part of the country, and are now in readiness to send for their supply six or eight ships; whereof his Majesty being advertised, we have received his royal command to signify his pleasure that you should represent these things to the States General, in his Majesty's name, who *jure primae occupationis,* (by right of first occupation,) hath good and sufficient title to these parts, and require of them that as well as those ships, as their further prosecution of that plantation may be presently stayed."

[1621] This remonstrance of the English Privy Council was made on the 15th of December, (O. S.) Sir Dudley Carleton appears to have delivered the remonstrance to the State of Holland the follow-

[1622] ing year. He informs the Council "that about four or five years previously, two companies of Amsterdam merchants began a trade to America, between 40° and 45°, to which they gave the name of New Netherlands, North and South Sea, &c., and had ever since continued to send vessels of 60 or 80 tons at most, to fetch furs, which is all their trade," and have factors trading with the savages; "but he cannot learn that any colony is as yet planted there, or intended to be." He, however, held an interview with the States, and presented a memorial dated Feb. 9th, of the subject of which they pretended to be ignorant, but promised on the 16th of March to write for information "to the participants of the trade in New Netherlands."[1] There is at present no written evidence of the result of this remonstrance to the Dutch, though a reply to it is informally referred to many years later. From this, as well as other evidence, it will be seen from the first, the right of the Dutch to the territories of New York, New Jersey and Delaware (which they claimed under the name of New Netherlands) was disputed by the English. It was never acknowledged, but constantly denied until their expulsion in 1664.

[1623] The West India Company having made arrangements to fulfill the objects of its charter, viz.: to trade with the natives and settle the country, fitted out a vessel called the New Netherlands, and appointed Captain Mey (from whom Cape May was named) and Adriaen Joriez Tienpont to be

[1] O'Callagan. London Docs.

directors of the expedition. Colonists, stores, provisions, and everything necessary were placed on board the vessels, and Mey and his companions took their departure for the Delaware, but which they called Zuydt *or* South, and also Prince Hendrick's River. They arrived safely and ascending the river for about fifteen leagues, immediately commenced the erection of a fort, which they named Fort Nassau. This was the first known building erected by civilized man on the banks of the Delaware. It was supposed to have been situated on the most northerly branch of Timber Creek, in New Jersey, not far from where the town of Gloucester now stands, a short distance below Philadelphia. On the map in Campanius'[1] work it is placed between the two branches of Timber Creek. But although the site is not certainly known, there is proof enough to show that it was within a short distance of Gloucester Point. We have no information as to how long Mey staid, or when he took his departure. It is supposed that he remained for some time, and carried on a trade with the natives for skins and furs, and that when he left he bore with him their affection and esteem. This, however, is all conjecture. Fort Nassau, after his departure, it is said by some writers, was abandoned, and the savages took possession of it. This was the case in 1633, when it was visited by De Vries, (who was the first that made an attempt at settlement within the limits of the State of Delaware,) and that it was then in the possession of a few savages, who wanted to barter

[1] The History of the Swedish Settlements on the Delaware.

furs.[1] But it must have been occupied the same year, as there are accounts of Arens Corson, a commissary, and a clerk residing there, who made purchases on the Schuylkill for the erection of another fort. As Van Twiller became Governor of New Amsterdam (now New York) this year, it is more than probable that hearing of its decayed condition, that he had it put in repair. In 1635 the English made an unsuccessful attack upon the fort.

[1624] Peter Minuit[2] or Minewa arrived at New Amsterdam as Director of New Netherlands. He became afterwards the Governor of Delaware. He generally receives the credit as being the first who ruled in the State, and his name is always placed in all publications first on the list of our chief magistrates. But Giles Osset, who administered the affairs of the Dutch Colony, massacred at Lewistown, was really our first Governor. Minuit, however, was the originator of the first permanent settlement both in our State, and on the banks of the river and bay, from which it derives its name.

The same year William Usselincx, a merchant of Antwerp, the original projector of the Dutch West India Company, presented a plan to Gustavus Adolphus, King of Sweden, for the formation of a Swedish West India Company, from which plan originated the settlement of the present State of Delaware. Usselincx, for some reason, became dissatified with the Dutch West India Company, of which he had for a long time been a director. He accordingly visited Stock-

[1] De Vries. [2] O'Callaghan.

holm, and proposed to the renowned Gustavus Adolphus, then the champion of the protestant interest in Europe, a plan for the organization of a trading company to extend its operations to Asia, Africa and America. Usselincx, in a written description, gave a glowing account of the advantages to be derived from this enterprise. In eloquent terms he represented such an establishment would be the means of planting among the heathens the Christian religion. That by it his Majesty's dominions would be greatly extended, his treasury enriched, his people's burdens at home diminished, and the nation not only relieved, but made prosperous by the establishment of a lucrative trade. The eloquent description of Usselincx had the desired effect. A company called the Swedish West India Company was formed [1626] and a charter granted them by Gustavus Adolphus. It was dated Stockholm, June 14, 1626, (old style). The following were its principal features.

Gustavus Adolphus (the King of Sweden) in granting it says:

"Finding it serviceable and necessary to the welfare and improvement of our kingdom and subjects that trade, produce and commerce should grow within our kingdom and dominions, and be furthered by all proper means, and having received of credible and experienced persons good information that in Africa, Asia, America and Magellanica, or Terra Australis, very rich lands and islands do exist, certain of which are peopled by a well governed nation, certain others by heathens and wild men, and others still uninhabited; and others not as yet perfectly discovered, and that not only with such places a great trade may be driven, but that the hope strengthens of bringing said people easily, through the setting on foot commercial intercourse, to a better civil state, and to the truth of the Christian religion," concluded "for the spread of the holy gospel and the prosperity of our subjects," to erect "a general company" or "united power of proprietors of our own realm, and such others as shall associate themselves with them, and help forward the work; promising to strengthen it with our succor and assistance, providing for, and founding it with the following privileges:"

The exclusive right for twelve years to trade beyond the Straits of Gibraltar, southward in the lands of Africa, and in America and Magellanica, or Terra Australis, reaching the coast of America at the like latitude as said straits, viz., 36°; also, with all lands and islands between Africa and America in same latitude. The vessels and goods of others than the company, who infringe those rights, to be confiscated. The government vessels of war, because not traders, to be exempted.

The company to be considered as commencing May 1, 1627, to continue for twelve years, during which none of the company have the power to withdraw the funds embarked in it, and no new members to be during that time admitted. If at the end of twelve years the company wish the term extended, it may be granted at the pleasure of the king.

Accounts are to be settled every year, at which

every person interested 1,000 scudis or thalers may be present. Every six years there shall be a final estimate of all accounts, and a new account begun. If then it appears to the majority of the stockholders that the profits or usefulness of the company do not justify its continuance, it may be dissolved.

Residents or landholders in Sweden may become members until the 1st of March, and those beyond the sea until the 1st of May next, after which none can enter the company either for small or large sums. The money to be paid in instalments, one-fourth on subscribing, and the remainder in three annual payments.

After the time for subscription shall expire, there shall be an election for regents or directors, in proportion to one for every 100,000 thalers subscribed. If, however, the subscribers of 100,000 thalers wishes it represented by two directors, it may be so, but the two only to receive the salary of one.

The directors to be chosen by a majority of the votes of stockholders, none to vote unless owning 1,000 thalers, and none to be a director who does not hold 2,000 thalers, and which sum while a manager, he cannot divest himself of.

The directors first chosen to continue in office for six years; after this, two-thirds to be newly elected, and one-third to be taken from the largest stockholders; this to be observed every two years until the expiration of the charter.

All countries, cities, and individuals who bring 100,000 thalers shall be entitled to appoint a director, and for this, all nations who have signed the agreement and transmitted the funds to some person in whom they may confide, and each individual subscribing shall declare the nation to whom he wishes to belong, and place his money.

Foreigners who decide to reside in Sweden, and contribute 25,000 thalers,[1] to enjoy the same privileges as citizens, and be free from every tribute, and as they carry on no trade, may depart at pleasure.

The directors to be all equal in power and authority, take oath of fidelity, administer justice without fear or affection, not deal in merchandise or own vessels.

They are to have a salary of 1,000 thalers per annum. In case of traveling for the company, besides their carriage, they shall receive six Swedish marks per day. The secretary and other servants to be paid out of the funds of the company; the directors of each chamber to be responsible for them.

If any damage result to the company from any of the directors, it shall attach to the chamber to which he belongs, and be refunded out of the funds contributed by it to the company.

Neither the directors nor their goods shall be liable for the company's debts.

All funds invested in the company shall be free from confiscation, even in the event of war of the King of Sweden with the nation of which the subscribers are a part.

Cities convenient for navigation, whose merchants

[1] About 74 cents.

contribute 30,000 thalers or scudi, shall constitute a chamber, or different cities or countries may unite their funds and agree upon the location of the chamber convenient to the company.

The company's vessels about departing from the different ports shall unite in a fleet at Gottenberg, and take their departure from thence, and at the end of the voyage return thither with their cargoes, which shall be unloaded, and thence transported, wind and weather permitting, without injury to the company.

If one chamber has goods, which another requires, they shall be furnished, so as to keep up a similar assortment in each.

There shall be one or more superintendents, who shall examine the accounts closely and consult with the directors on important matters connected with the interests of the company, and in elections of superintendents, captains, &c., required, stockholders shall be preferred, if equally capable.

Superintendents may be removed from one chamber to another, and every chamber shall have a representative at Gottenberg, and be informed within two months after the sailing of the vessels of the matters connected with the voyages, and every three months furnished with an account of goods sold.

When necessary there shall be held a diet or meeting of all the chambers, to take place alternately, at different chambers, in the order of the largest subscriptions, the object being to discuss all the general interests of the company, voyages, freights, prices, &c.

To each diet, twelve managers shall be sent from

each chamber, and the government to be entitled to one vote, making thirteen, or casting vote; every chamber having a vote in proportion to its furnished capital; a chamber furnishing half has six votes, one third four votes, and a majority to decide.

On all imports or exports to or from Sweden, a duty to be paid of four florins per cent., which payment entitles them to be transported freely thereafter through the whole kingdom.

The company to be under the royal protection, in the free exercise of its trade, the use of its vessels, and defence against all attempts to injure it in war or in peace.

The government to furnish vessels of war, forts, soldiers, guns, &c., at its own expense. All vessels, &c., taken by the company from pirates, &c., shall be for the company's benefit, except where they are assisted by the government vessels, in which cases the prizes to be divided equally.

The government not to use the vessels of the company, nor their funds or merchandise, even in war, without its consent.

The company shall be entirely at liberty, within the aforesaid limits, to make treaties with foreign chiefs or people in their own name; to build cities, castles, fortresses; occupy desolate places, and make them habitable; operate and procure what they can, of use to, and for the convenience of the company; but not to commit violent hostilities against the inhabitants of the country, nor, unless so tempted, do any other thing against the subjects of the King of Spain, nor exercise commerce in places with their subjects without their express license, under pain of penalties against transgressors of the King's orders and disturbers of the public peace.

In case of ill treatment in the use of its trade, or by force or fraud, the company is at full liberty to avenge itself on its enemies as against pirates and robbers, &c.

In order to manifest the desire of the government to aid and improve the company, it will contribute and put at equal risk with others, 400,000 Swedish dollars.

The government, besides the four florin per cent. duty, will receive one fifth of ores, silver and other minerals, which may be transported from the mines, and one tenth of the fruits of the country, in recompense for its aid, privileges, &c., granted. The merchandise, and metals received from merchandise, to be exempted and remain for the country. "And, whereas, William Usselincx, of Brabant, Antwerp, has spent much time of his life in seeking out said ports, and by the testimony of the State of Flanders, and Maurice, Prince of Orange, he is stated as the chief inventor in Holland of the West India Company, and by him, its administration has been much aided, and having already resolved to establish in Sweden, has promised faithfully to exert himself; therefore, to recompense him, the Company are to pay him one florin per 1000, of the merchandise which the company shall import or export during its traffic within the limits of its charter."

The company to constitute a council, which, with its officers, shall attend to the administration of justice, preservation of good laws, continuation of war; appoint soldiers, governors, directors and judges; build castles and cities; accommodate differences between the citizens of the country and the natives, as well as between directors or chambers, and finally preserve everything in good condition, and under good order.

This council must consist of the chief stockholders, and attend to the business and consignments on commission and others, furnish information of the ships and advices received, and decide on operations. The number of council to be determined by circumstances and the judgment of the company.

If any chief community, city, or company, contribute 500,000 to the company, it may appoint an agent with full powers to negotiate about things necessary to be done.

If the company requires alterations in the conditions and of the charter, not contrary to the laws and welfare of the republic, they may be conceded to it.[1]

The eloquent description of Usselincx, and the granting of the charter, created a perfect furore amongst all ranks in Sweden. The historian of the

[1] Hazard's Annals. It was translated for this work from, "Aurgonautica Gustaviana," printed in Frankford, 1633, a very rare work, the only copy known to be in this country is in the valuable library of Harvard College, to whose librarian Hazzard was indebted for the use of the work. It is in the German language. The charter is also to be found in the Italian language, in the fourth edition of Harte's Life of Gustavus Adolphus.

life of Gustavus[1] says, "It is not to be described, how much all these new schemes delighted the Senators, particularly that relative to the establishment of the West Indies (as America was then called), to which all people subscribed readily and generously, in conformity to the example set them by the king." Another writer[2] says that the plan was supported by the king's mother, by Jno. Cassimer, Prince Palatine of the Rhine, who had married the king's sister, by the members of his majesty's councils, by the principal nobles, general officers, bishops, clergy, burgomasters, councillors of cities, and the greatest part of the commonalty, and that a time was appointed for bringing in the amount subscribed in Sweden proper, in Finland, Sivonia, and elsewhere. The "ships and all necessaries were provided; an admiral, vice-admiral, officers and troops, commissaries, and merchants, and assistants were appointed. The work was ripe for execution, when the German war, and afterward the king's death prevented it, and rendered the fair prospect fruitless." Others assert, that a squadron was fitted out and sailed for America,[3] but this is not well authenticated. Campanius asserts that "the designs of Gustavus could not be carried into full effect, because he was engaged in a war with six powerful enemies, and because *the ships for that purpose were stopped and detained by the Spaniards in their voyage* (*to America*), *which was done in order to favor the Poles and emperor of Germany, then engaged*

[1] Harte. [2] Campanius.

[3] Campanius and Harte, Life of Gustavus Adolphus.

in a war against us." But documentary evidence shows great doubt on this. Campanius was an unreliable writer. And the received opinion amongst historians is, that no attempt was at that time made by the Swedes to settle America. But, that the purpose then formed, was afterwards carried into effect, some twelve years later, viz., in 1638, when the first permanent settlement on the Delaware was formed within the present limits of the city of Wilmington. The archives at Stockholm showed that preparations were at that time made to settle this country, but did not record the failure, and hence, the error of many early historians, in their relation of the early settlements on the banks of the Delaware. Gustavus Adolphus, to whom, in no small degree is owing the first settlement of the State, was afterward killed at the battle of Lutzen, in the year 1632. The expedition that finally did make permanent lodgment in Delaware, did not sail until five years afterwards, in the reign of his daughter Christina, who was born the 9th day of December (O. S.), or about six months after the grant of the charter, for the settlement on the banks of the Creek, which long bore her name, but which was afterward corrupted into Christiana.

CHAPTER IX.

FROM 1629 TO 1633.

Delaware a part of the Dutch Colony of New Netherlands—Charter of the Assembly of XIX to settle New Netherlands—Features of the Charter Grant to Herr Samuel Godyn of lands in South (Delaware) River—First Land Grant in Delaware—Sale of land by the Indians, to Godyn and Blommaert—First deed in Delaware—Godyn and Blommaert form a partnership to settle this State—Peterson De Vries—Sailing of a Colony from Holland for Delaware under De Vries—They settle at Hoornkill (Lewistown)—They build Fort Oplandt—Name the place Swanendale—De Vries leaves for Holland—Delaware Bay called Godyn's Bay—Massacre of the Settlers by the Indians—Arrival of De Vries in the Delaware—Interview with the Indians—Their account of the Massacre—Asserted relinquishment of the Delaware by the English to the Swedish Ambassador—Governor Minuit recalled to Holland—Grant to Lord Baltimore—His death—Death of Gustavus Adolphus, through whose influence Delaware was first settled.

[1629]

For the purpose of promoting the settlement of what the Dutch called New Netherlands, then consisting (according to the Dutch claim) of the present State of Delaware, New Jersey, New York and part of Connecticut; the Assembly of XIX, on the 7th of June, granted a charter of "Freedoms and Exemptions, to all such as shall plant colonies in New Netherlands." This gave the privilege to members of the company to send to New Netherlands in the company's ships, on certain terms, three or four persons to view the country, for the purpose of selecting lands. Those, who after four years notice to the company, planted a colony of fifty souls, over fifteen years of age, were to be entitled to the dignity of being made patroons. They were allowed to have the privilege of selecting lands for four Dutch miles (about sixteen English) along the shore on one side of any navigable river, and as far into the interior of the country as their situation would permit. If they selected their lands on both sides of a navigable river, they had only two Dutch or eight English miles in length. The company reserved the right to the land between the limits of the colonies to themselves, under the general rule, that no person should be allowed to come within thirty-two English miles of them without their consent. The jurisdiction of the river was reserved to the States General or Company. The patroons were to enjoy and possess over the lands within their limits, fruits, rights, minerals, rivers, and fountains; have "chief command and lower jurisdiction," fishing, fowling, and grinding, exclusively. They had also the privilege of founding cities, appointing officers and magistrates, besides other powers and privileges.[1] It was, in fact, transplanting the feudal system of Europe to the shores of America. Under this grant, "The Herr Samuel Godyn (a merchant of Amsterdam), and Samuel Blommaert, on the 19th of June, obtained a grant of land on the west side of South River Bay, extending from Cape Henlopen inland thirty-two miles, and two miles in breadth. They had pre-

[1] O'Callighan, N. Y. Documents.

viously sent persons to examine it, and purchased it from the Indians. This was the first grant of land made to any European in the State of Delaware, or on the banks of the river and bay of that name. An Indian village then stood somewhere in the neighborhood of Lewistown, possibly on the ground on which that town now stands. Moulton speaking of this purchase says, "One of three ships sent over by the department of the West India Company, this year (1629), visited the Indian village on the Southwest corner of Newport, May, or Delaware Bay, and that the purchase was then made from Cape Hindlop to the mouth of the river." What river we are left to infer. But as in the deed (in Moulton) speaks of the extent of the grant being eight large miles, and as the Dutch mile measures in length that of four English miles, therefore the extent of the Indian grant to the Dutch would be thirty-two English miles along the coast of our State from Cape Henlopen northwards.

The river alluded to therefore, must either be Jones' or Murderkill Creeks, or Mahon River. It is more than probable the latter. For as the Dutch (judging from the usual course of the dealings of the white man with the Indian) would be more apt to over than under measure their purchase, we may safely judge that the Mahon, which is over thirty-eight English miles as the crow flies, from Cape Henlopen, would be more likely to be the one alluded to, than the Jones or Murderkill, which are hardly twenty-six miles.

This grant therefore, comprised nearly the whole

bay front of Kent and Sussex counties. The deeds of this land has been happily preserved in the New York State Library, and as it is the first deed ever given for land in Delaware, or on the banks of the Delaware we have published it entire. It is as follows:

"We, the Directors and Council of New Netherlands, residing on the Island of Manhattan and in Fort Amsterdam, under the authority of their High Mightinesses the Lord's State General of the United Netherlands, and of the Incorporated West India Company Chamber at Amsterdam, hereby acknowledge and declare, that on this day, the date underwritten, came and appeared before us in their proper persons, Queskacous and Entquet, Siconesius and the inhabitants of the village, situate at the South Cape of the bay of South River, and freely and voluntarily declared by special authority of the rulers, and consent of the commonality there, that they already on the first day of June, of the past year 1629, for, and on account of certain parcels of cargoes, which they previous to the passing hereof, acknowledged to have received and got into their hands and power, to their full satisfaction, have transferred, ceded, given over, and conveyed, in just, true, and free property, as they hereby transport, cede, give over, and convey to and for the behoof of Messrs. Samuel Godyn and Samuel Blommaert absent; and for whom, We, by virtue of our office under proper stipulation, do accept the same, namely, the land to them belonging, situate on the south side of the aforesaid Bay, by us called the Bay of the South

River, extending in length from Cape Hinloffin, off into the mouth of the aforesaid South River, about eight leagues (groote mylen), and half a league in breadth into the interior, extending to a *certain marsh* (lieyte) *or valley*, through which these limits can clearly enough be distinguished. And, that with all the action, right, and jurisdiction, to them in the aforesaid quality therein appertaining, constituting and surrogating the said Messrs. Godyng and Blommaert, in their stead, state, zeal, and actual possession thereof; and giving them at the same time, full and irrevocable authority, power, and special command to hold in quiet possession, occupancy and use, tanquam Actores et Procuratores in rem propriam the aforesaid land, acquired by the above mentioned Messrs. Godyn and Blommaert, or those who may hereafter obtain their interest; also, to so barter and dispose thereof, as they may do with their own well and lawfully acquired lands. Without the grantors having reserving or retaining for the future, any of the smallest part, action, right, or authority, whether of property command or jurisdiction therein; but now, hereby forever and a day, desisting, retiring from and abandoning, and renouncing the same, for the behoof aforesaid, promising further, not only to observe, fulfill, and hold fast, steadfast and unbroken, and irrevocable, that their conveyance and whatever may be done in virtue thereof, but also the said parcel of land to maintain against every one, and to deliver free of controversies, gainsays, and contradictions, by whomsoever instituted against the same. All in good faith, without guile, and deceit. In witness this confirmed with our usual signatures, and with our seal dependant therefrom. Done at the aforesaid Island, Manhattan, this 15th July, xvi. and thirty.[1]

PETER MINUIT, *Director*,
JACOB ELBERTSON WISSINK,
JAN JANSEN BROUWER,
SIMON DIRCKSEN POO,
REYNER HARMENSEAR,
JAN LAMPE, *Sheriff*."

It would be impossible at this day, to find out the grant by these landmarks, which the deed says, "*through which by these limits can clearly enough be distinguished*." The valley does not now (if it ever did) exist in Kent county. If the changes in the country by the encroachments of the bay has not destroyed the hills, the landmarks the deed alludes to, must have been in New Castle county. The term marsh is very indefinite, as with but few intervals of fast land (such as at Kett's Hammock, in Dover hundred, Bower's Beach, in South Murderkill hundred, Kent county, and Thorn Point, in Cedar Creek hundred, Sussex county), the whole bay coast between Mahon River and Lewistown, is marsh.

[1] A photographed copy of the original of this deed, was presented to the Historical Society of Delaware, by General Meredith Reed. This is the first time it was ever published. This deed, as given by Moulton, is published in Hazard's Annals, page 23. The name, however, of Blommaert, is not inserted in that deed. There are no signatures to it, and the name of the Indian grantors are given as Queskakous, Esanques, and Sickonesgris.

After the grant of this land to Godyn and Blommaert, they formed a partnership with several others to attempt a settlement, also with a view of engaging in the whale fishery, "whales being plenty in those regions, and fish oil being 60 guilders per hogshead" in Holland. Previous to forming this partnership, however, he met with *David Pieterszen De Vries*, of Hoorn, a port in North Holland, "a bold and skilful seaman, and master of artillery in the service of the United Provinces." He had about two months previously returned from the East Indies. An offer of a "commandership" was made to him by Godyn, or Blommaert, or both, and employment as "second patroon," such as granted by the State, and by the 19th Article of the West India Company's Charter. This he declined to accept, unless he was made equal in all respects to the others as patroon, which, being readily agreed to, a patroonship was formed by entering into formal articles of association on the 16th of October. Those who composed it were Samuel Godyn, William Van Rensselaer, Samuel Blommaert, Jan De Laet, and De Vries, to which several others were afterwards added. Preparations were made immediately for the expedition, a ship and yacht were fitted out, thirty colonists placed on board, with material for whaling and for planting tobacco and grain, and thus equipped, on the 12th of December, under command of De Vries, they sailed from the Texel, to make the first attempt at settlement in the State of Delaware.[1]

[1] De Vries, N. Y. His. Collection.

[1631] When De Vries' expedition arrived in the Delaware is not certainly known, but as he sailed in December, taking the usual length of passages at that time as a criterion, he must have arrived in the Delaware Bay in the March or April of the following year. After passing Cape Cornelis, he entered a deep creek, abounding with oysters, which he named Hoornkill or Hoorkill, probably after Hoorn, the place of his residence in Holland, and kill, the Dutch name for creek. In other words, Hoorn Creek, afterwards called Whorekill.[1] It is the present Lewes Creek in Sussex county. Here he erected a house, and surrounded it with palisades instead of parapets and breastworks, which served the purpose of both trade and defence. He named it Fort Oplandt. The weather was fine, and no inconvenience was suffered from it. From the number of swans which he had seen, he named the place Swanendale, or "Valley of Swans." De Vries sailed some time in the course of the year for Holland, leaving the colony in command of Gillis Hossett, the commissary of the expedition. Either before or after the departure of De Vries, a purchase was made by Gillis Hossett from the Indians, for Godyn and Blommaert, for a tract of land, on the east side of Delaware Bay, or Cape May, in the present State of New Jersey. Both sides of the

[1] Several writers have said that this name was given from the bad conduct of the Indian women. But there is no just reason for this statement. The Dutch always called it Hoornkill. It was not until after the arrival of the English that it was called Whorekill. Until the arrival of Penn, Whorekill was the name given to the whole of Sussex county.

river were now named Swanendale. The purchase was made on board the "Ship Walrus," or Whale, before Peter Heysen, skipper, and Gillis Hossett, commissary. This ship Walrus was probably the yacht that came over with De Vries for the purpose of whaling. Hossett, who may be called the first *governor of Delaware,* as he governed the colony of Swanendale, had formerly been agent for the purchase of lands around Fort Orange (now Albany, N. Y.) for Van Rensselaer. The bay was at this time called Godyn's Bay.

Some time after the departure of De Vries, this unfortunate colony (the first settlers of our State) were all massacred by the Indians. The setttlers under Mey had at this time abandoned Fort Nassau, and the only white residents on the Delaware were the colonists at Swanendale. The account, as learned by De Vries on his second visit, was as follows:

The Dutch, according to their custom, had erected a pillar, on which was a piece of tin, on which was traced the coat of arms of the United Provinces. One of the chiefs wanted to make it into tobacco pipes, and not knowing that it was improper, took away the tin, which gave the officers in command much dissatisfaction, so that the Indians did not know how to make amends. They went away and killed the chief who had taken the tin, and brought a token of it to those who commanded at the house, who told them that they had done wrong; that they ought to have come with him to the house, and they would have told him not to do so any more. They then left, but the friends of the murdered man resolved to be revenged. They attacked the Dutch when they were working in the field, leaving but a single sick man in the house, and a large bull dog, which was chained out of doors. The man who had command of the house stood near the door. Three of the boldest Indians who were to perpetrate the deed, came and offered him a parcel of beavers to barter, and contrived to enter the house. He went in with them to transact the business; that being done, he went to the garret where the stores were. Coming down, one of the Indians cleaved his head with an axe, so that he dropped dead on the floor. They then murdered the sick man, and then went to the dog, which they feared most, and shot at least twenty-five arrows at him, before they killed him. They then went in a treacherous manner to the people in the field, approaching them with the appearance of friendship, and murdered one after another. "Thus," says De Vries, "terminated our first colony, to our great loss."[1] Thus also perished the first white inhabitants of the State of Delaware.

[1632] De Vries by some means had heard of the destruction of his colony before he left the Texel, which was in the latter part of the year. The news had more than probable been carried by the Walrus (which appears to have been an appendage to the colony for the purpose of whale fishing) or some other vessel, to New Amsterdam, and from there to Holland. Long before he saw the land, he knew he

[1] De Vries, N. Y. His. Soc. Coll., vol. 1, N. S., p. 52.

was near the coast, "by the odor of the underwood, which at this time of the year is burned by the Indians, in order to be less hindered in their hunting." On the third of December he saw the entrance of the bay, on the 5th sailed within the cape, having a whale near the vessel, and on the 6th ran with the boat up the Hoornkill, having first put themselves in a proper state of defence in case of a hostile attack from the Indians. They found their dwelling house and store had been burnt to the ground, and their fortification utterly destroyed. The ground was bestrewed with the heads and bones of the murdered men, but he saw no Indians. Supposing that they might be attracted by the sound of a gun, he went on board the vessel and ordered the guns to be fired. On the 7th the Indians appeared near the destroyed house; afraid to approach, they wished the people from the vessel to come on shore, which De Vries resolved to do next day, in the yacht, that he might "have a shelter from their arrows." Accordingly he went in the yacht up the creek to the house. The Indians were on the shore, but at first they would not go on board; at last, however, one entered the vessel. De Vries gave him a cloth dress, and told him he desired to make peace with them. Others then went on board, expecting also a dress, but he gave them only trinkets, adding that the dress was given to the first as a reward for his confidence in venturing to enter the boat. They were desired to come on board with their chief, called Sakimas, for the purpose of making a satisfactory peace. One

Indian, however, remained on board all night in the yacht, and from him De Vries, upon inquiry, received the account which has already been given of the murder of the colonists. On the 9th the Indians, with their chief, came. They sat down in a circle, and concluded peace. Presents were made them of duffels, bullets, axes, and Nuremberg trinkets, with which being well pleased, and with promises of reciprocal benefits, they departed joyfully, no vengeance having been taken for the previous cruelties. De Vries now made preparations for the fishery, and for boiling oil, by forming a lodging place of some boards.

About this time it is said that upon the application of John Oxensteirn, the Swedish Ambassador, King Charles I. relinquished to the Swedes all claims to to this part of the country by reason of discovery. There is no documentary evidence to support this agreement. Acrelius refers to this circumstance, but places it at a later period.[1]

During this year Director Minuit, who afterwards commenced the settlement at Fort Christina, and was the first governor of Delaware after its permanent settlement, (or by any portion of the ancestors of the present people of this State,) was recalled to Holland. He embarked from New Amsterdam in the spring.

On the 20th of June, O. S., Cecilius, Lord Baltimore received the grant for the present State of Maryland. His intention was to settle in Virginia, but being a Catholic, and finding himself uncomfortable on account

[1] Hazard's Annals.

of his religion, he departed for England, and obtained a patent for the land between that of the North and South Virginia companies. There he died before his return. The next year his son had his patent confirmed to himself. The terms of his grant included not only the present State of Maryland, but the whole of Delaware, and that part of the State of Pennsylvania from the Delaware line a short distance north of Philadelphia, including in its limits the present city of Philadelphia, and a great proportion of the counties of Chester, Delaware, Lancaster, York, Adams, Franklin, Fulton, Bedford, and Somerset. The following description will show the extent of the grant:[1]

"By letters patent of this date reciting the petition of Cecilius, Lord Baltimore, for a certain country thereinafter described, *not then cultivated and planted, though in some parts thereof inhabited by a certain barbarous people, having no knowledge of Almighty God*, his majesty granted to said Lord Baltimore all that part of a peninsula lying in the parts of America between the ocean on the east and the bay of Chesapeake on the west, and divided from the other part thereof by a right line drawn from the promontory or cape of land called Watkins' Point (situate in the aforesaid bay, near the river of Highco) on the west, unto the main ocean on the east, and between that bound on the south, unto the part of Delaware Bay on the north which lieth under the 40th degree of north latitude, from the equinoctial where New

[1] Beverly, 47, 48; Barkes' Virginia, 11, 39; Bosman's Maryland.

England ends; and all that tract of land between the bounds aforesaid, *i. e.* passing from the aforesaid bay, called Delaware Bay, in a right line by the degrees aforesaid promontory, or place called Watkins' Point."

This grant was a consequence of many disputes, both with the Dutch, the Duke of York, and Penn. Under it Baltimore claimed possession of Delaware. This matter will be found treated of more at length in another portion of this history.

Gustavus Adolphus lost his life at the battle of Lutzen, the 16th of October. A short time before his death, whilst at Nuremberg, he drew up a scheme of a company, which was ready for his signature, but which was prevented by his death. It was however the next year submitted to the people by his chancellor, Oxensteirn. It was in a great degree to the exertion of this renowned prince that Delaware owed her first permanent settlement.[1]

[1] Statement of Penn's cases, by Murray. Reg. Penn., vol. 2, p. 204.

CHAPTER X.

FROM 1633 TO 1637.

De Vries sails up the Delaware to Fort Nassau—He is warned by an Indian woman of treachery—Massacre of English boat's crew—Indians warned to go on shore—They make peace with De Vries—De Vries sails for Virginia—Is informed the English claim South River—Is told the murdered boat's crew belonged to Virginia—De Vries returns to South River—Bad success of whale fishing—He returns to Europe—War between the Timber Creek Indians and the Minquas—Publication of the Charter of the Swedish West India Company by Chancellor Oxensteirn—Grant to Sir Edward Plowden by Charles I.—Sale of the colony of Swanendale to the Dutch West India Company.

De Vries, whom we made mention in the previous chapter as having concluded a peace [1633] with the Indians, remained some time in the neighborhood of Lewes Creek, during which interval it is supposed he engaged in whale fishing, proceeded in his yacht up the Delaware, to procure beans from the Indians. At the mouth of the river he saw a whale. When opposite Fort Nassau, he found a few Indians disposed to barter some furs, but wanting only Indian corn, and having disposed of most of his articles at Swanendale, he had none to trade for furs. The Indians advised him to proceed to Timmerkill (now Timber Creek). But he was prevented by the kind interference of an Indian woman, to whom he had given a cloth dress to induce her to communicate what she knew. She belonged to the "Sankitans." She informed De Vries that they had murdered the crew of an English boat that had ascended the Count Earnest (Delaware) river and would undoubtedly attack them. On the 6th, he weighed and stood before Timmerkill (Timber Creek) fully prepared for the Indians if they intended to harm him. They soon approached the boat, and about forty-two or forty-three entered the yacht. Some began to play on reeds, so as to induce no suspicion of their designs, but, being only seven in number, the crew were upon their guard, and when De Vries thought they had been long enough on board, he ordered them ashore, threatening to fire if they refused to depart. The sachem offered beavers for sale, which were declined, but the Indians were again ordered on shore, and given to understand that Manito, their devil, had advised them of their evil designs. They then went on shore.

These Indians were said to be "Roodehoeks or Mantes." They were partly dressed in English jackets, which created suspicion, and confirmed the story of the Indian woman. On the 8th, De Vries returned to his position before the fort, which was now crowded with Indians, and their numbers increasing. A canoe with nine chiefs from different places came off, amongst them the man who had appeared with the English jacket, which, however, he did not now wear. They sat down in a circle, and said they had discovered that De Vries' people were in fear of them, but they came to conclude a permanent peace, and presented ten beaver skins, with a distinct ceremony with each. De Vries, upon the receipt of each article, offered some presents, such as axes, adzes, or small knives, &c., which they refused, saying "they did not make these presents to receive others in return, but to make peace." They replied, "these must be given them on shore." De Vries, on the 9th and 10th obtained from them in barter some Indian corn and furs.

De Vries failing to obtain corn in the South (Delaware) River, sailed for Virginia. Upon his arrival there, he was met by the governor, attended by some officers and soldiers, who gave him a cordial welcome. Upon inquiring from whence he came, and being informed from South River, the Governor invited him to his house, and treated him to a glass of wine. He then told him that South River belonged to the British, and was by them named Delaware Bay, after Lord Delaware, who some years ago had taken possession of it, but not supposing it navigable owing to the sand banks, he did not ascend the river. De Vries informed him he was mistaken. That the Dutch had built a fort there many years ago, called Nassau, of which he appeared never to have heard before, and that it was a fine navigable river. The governor spoke of a small vessel that he had sent, some time before, to the Delaware, which had not returned, and he supposed was lost. De Vries then related the circumstance mentioned by the Indian woman of the murder of an English boat's crew, and that he had seen an Indian wearing an English jacket, which he had concluded belonged to his boat's crew, which had been sent there to make discoveries. The governor gave him six goats for his new colony; he then purchased some provisions, and returned to South Bay. He there learned that in his absence they had only taken seven whales, which yielded 32 cartels of oil. Finding that the fishing here was too expensive in proportion to the profit, and the fish poor, he returned to Fort Amsterdam, and from there to Europe."[1]

The Timber Creek Indians at this time were at war with the Minquas on the Christiana. The name of the sachem belonging to the former tribe was Zuee Pentor.[2]

The Chancellor Oxensteirn, on the 10th of April, published the proclamation which had been left unsigned by Gustavus Adolphus. The chancellor also added:

"Though the above declaration and amplification of the before mentioned privileges of his majesty of glorious memory could not have been signed on account of the multifarious and incredible affairs of the war, I cannot, in consequence of my duty and good personal knowledge, but certify that the same has been the highest desire and wish of his royal majesty; therefore I, by the crown of Sweden, and plenipotentiary minister general, have signed it with my own hand, and affixed my seal to it at Hilebrum, April 10th, 1633."

The chancellor also published an address, in which

[1] De Vries, N. Y. His. Coll. [2] De Vries.

he asserted that it was the desire of Gustavus Adolphus that there should be a "general commercial and navigation company" in Sweden, and that "free and open commerce should be carried on in every part of the world where the greatest advantages might be derived." He also stated that "the work was almost carried into operation," but was delayed by the absence of the king in the crusades, in Prussia and Germany, and from other causes. He appointed as first director of the company, W. Usselincx, the indefatigable Antwerp merchant, through whose exertions it was the first settlement was made in this State.

[1634] On the 21st of June, a patent was granted by King Charles the 1st to Sir Edward Plowden. It is said under this grant that a settlement was made on the banks of the Delaware. Many writers suppose it was within the limits of this State. Huffington was of this opinion.[1] But the courses and distances in the grant do not seem to apply to Delaware.

The grant to Plowden was from "Cape May, and from thence to the westward for the space of forty leagues, running by the river Delaware, and closely following its course by north latitude, unto a certain rivulet there, arising from a spring of the Lord Baltimore and the lands of Maryland, when it touches, joins and determines in all its breadth, from thence takes its course to a square leading to the north by a right line for forty leagues, &c.; thence

[1] Huffington's Delaware Register.

likewise by a square inclining to east in a right line for the space of forty leagues," and from thence by various other courses and distances, mentioned in the grant, to the place of beginning, at Cape May. This region was called New Albion. But as it runs by the course of the Delaware, and as following the course of that river from Cape May, would lead in a northwest direction, the grant to Plowden, was from all the evidence that can be procured, from Cape May to about Trenton, from there to the neighborhood of Jersey City, opposite Long Island, and including that island to Cape May again. Thus this grant of Plowden appertained exclusively to New Jersey,[1] and had no relation to Delaware.

[1635] On the 7th of February the two colonies of Swanendale, one heretofore described, comprising that portion of Kent and Sussex which fronts on the bay, the other on the opposite coast of New Jersey, were sold by the patroons to the West India Company for 15,600 guilders, or $6,240. This was the first land sale made by white men in the State. There appears to have been a difficulty between the patroons and the company, as in the grant it was stipulated that "the right of neither party in a suit depending between the patroons and the company at Amsterdam was to be impaired." The purchase-money was to be paid in installments on the 27th day of each of the months of May, August, and November, 1635. The outstanding

[1] In 1648 a pamphlet was published, giving a description of New Albion. It may be seen at length in Smith's History of New Jersey.

accounts between the settlers and the Company to be considered as discharged, and they not subject to duties on an export cargo of timber.[1]

No other event is recorded in any manner relating to the State until 1637.

[1] See the agreement at length in O'Calligan.

CHAPTER XI.

FROM 1637 TO 1638.

Peter Minuit—Embarkation of the Swedes from Gottenberg for Delaware, under Minuit—Voyage—Arrival at Jamestown, Va.—Letter of Governor of Virginia in relation thereto—English claim the country—They arrive at the Delaware—They land between Murderkill and Mispillion Creeks—Name it Paradise Point—Sail up the Delaware, enter Minquas Creek, and land at the Rocks—Appearance of the Country—Build Fort Christina—Change the name of Minquas to Christina Creek—Build Christnaham—Purchase land from Indians—Dutch object to Swedish settlement—Protest of Director Kieft—Cruelty of Kieft to Indians—Purchase from Cape Henlopen to Trenton—Review of rights of Dutch and Swedes—Remarks thereon.

[1637] THIS year the energy of Usselincx, the projector both of the Dutch and Swedish East India Companies, to which is owing both the settlement at Manhattan (New York) and the South River or New Swedeland stream (the Delaware) as it was named by the Swedes, bore fruit. Gustavus Adolphus, the renowned King of Sweden, had been killed five years before at the battle of Lutzen, and his infant daughter, Christina, was now seated on the throne of Sweden. Peter Minuit, who had been appointed as Director-General of New Netherlands, in 1624, and who had been recalled in 1632, had quarreled with the company who had employed him, and offered his services to the crown of Sweden. The

Dutch West India Company had been engaged in disputes with the patroons, or large landholders possessing feudal powers, and their interest was opposed to the monopoly of trade enjoyed by the Company. Minuit appears to have been a man of integrity, who endeavored to maintain the legal right of the company, or in other words, their monopoly of the commerce between New Netherlands (as the Dutch possessions in this continent were called) and the mother country. He was undermined, by interested statements made to the directors, and probably smarting under what he considered their injustice, laid before the celebrated Chancellor Oxenstiern a plan for the settlement on the Delaware, and offered to conduct the enterprise. His offer was accepted. He was appointed governor of the expedition, and accordingly some time in the fall of the year it is supposed (for the true date of the sailing of this expedition has never been ascertained), he set sail from the port of Gottenberg on the west coast of Sweden. The expedition consisted of two vessels. One was an armed ship, called the "Key of Kalmar," named after a town in Sweden, the other a transport ship, named the "Bird Grip," or Griffin. The expedition, it is supposed, numbered about fifty persons, many of whom it is said were criminals, as it was the custom of the Swedish as well as other governments at that time to transport convicts and laborers to the colonies, where they were sold as indented servants.[1] They were well supplied with provisions for the colony, and

[1] Acrelius, 408.

arms and ammunition for defence, with merchandise for trade, and with presents for the Indians. They also brought over with them Reorus Torkillus, a clergyman, who died in 1643. This was the first clergyman that ever preached on the banks of the Delaware. He was buried, it is supposed, in the graveyard of the old Swedes Church in Wilmington. They [1638] sailed by the way of the West Indies. On their route they touched at Jamestown, Virginia, and made known to the English authorities there that they were bound to the South River. The English desired a copy of their commission, but this was refused, unless they were allowed free trade in tobacco to carry to Sweden. This was not complied with, as being contrary to the English king's instructions. The vessel remained at Jamestown for about ten days, to refresh with wood and water, and then proceeded on their destination to the Delaware,[1] where

[1] The following letter from Jerome Hawley, Secretary of Virginia, to Mr. Secretary Windebanke, gives an account of this visit of the Swedes. It is copied from Broadhead's London Documents, at Albany, vol. 1, pp. 57 and 58:

"Jamestown, in Virginia, *May* 8, 1638.

"Right Hon.—Upon the 20th of March last I took the boldness to present you with my letters, wherein I gave only a touch of the business of our Assembly, referring your honor to the general letters then sent by Mr. Kemp, from the governor and Council. Since which time have arrived a Dutch ship, with commission from the young Queen of Sweden, and signed by eight of the chief lords of Sweden, the copy whereof I would have taken to send to your honor, but the captain would not permit me to take any copy thereof, except he might have free trade for to carry to Sweden, which being contrary to his majesty's instructions, the governor excused himself thereof. The ship remained here about ten days, to refresh with wood and

they arrived in the month of April of the same year. As the expedition entered our beautiful Bay at the time all nature was shining forth in verdure, they came to a point of land jutting into the broad expanse of its waters, on which they landed for observation and refreshment. It was situated between the present Murderkill and Mispillion Creeks, in what is at present Kent county. They were so struck with its beauty that they named it Paradise Point. It was only a few miles from where De Vries' colony had been massacred by the Indians.

After refreshing themselves a brief time at Paradise Point, they weighed anchor and sailed up the Delaware, and passing the point where New Castle now stands, at the distance of four miles above it, they found the land on the left to trend away towards the west and northwest, forming a cove about three miles long, and varying in width from one furlong to one or two miles. They sailed on until they entered Minquas Creek, (the Christiana,) and proceeding up it for about two and a half miles, cast anchor at the

water, during which time the master of said ship made known that both himself and another ship of his company were bound for Delaware Bay, which is the confines of Virginia and New England, and there they pretend to make a plantation, and to plant tobacco, which the Dutch do so already in Hudson's River, which is the very next river northward from Delaware Bay. All which being his majesty's territories, I humbly offer the consideration thereof unto your honor, and if his majesty should be pleased to think upon any course, either for removing them, or preventing others from settling upon his majesty's territories, I humbly conceive it may be done by his majesty's subjects of these parts, making use only of some English ships that resort hither for trade yearly, and be no charge at all upon his majesty."

Rocks. These rocks form a natural wharf of stone, and are situated at the foot of Sixth street in the city of Wilmington, and the site of which was until recently occupied as a ship yard, and at that time were one of the capes of the Christiana. The other was the old Ferry Point, where the Townsend Iron Works now stand. The marshes on the south and east on both sides of the Christiana, on the south and east of the city of Wilmington, known as the Cherry Island, Holland's, Middleborough, and Deer Creek marshes were then at high tide under water, save a small island known as Cherry Island, which gave the former marsh its name. Therefore between the Rocks, the old Ferry Point at high tide, and the the Jersey shore opposite, was a waste of waters, the Delaware then washing their shores. These rocks are the termination of a vein of hard blue rocks, which "issue from our loftiest hills." This ledge, passing across the country in a southerly direction, sometimes dipping beneath the surface, sometimes just showing itself above it, at length arrive at the shores of the Brandywine, a short distance below the mills. There it presents a high, bold point, and then sinks beneath the channel of that river. After passing under it, and a narrow strip of meadow land on its south side, it immediately rises in large naked masses, and proceeding onwardly, mostly below the soil, at length terminates abruptly on the margin of the Christiana,[1] and forms the rocks, the natural wharf above described. The land covering this ledge from the point

[1] Ferris' Original Settlements.

of rocks toward the north rises with a gentle, continual swell, widening as it recedes from the Christiana, and standing high above the meadows on either side. It extends in that direction about six hundred yards, and then gently declines towards the Brandywine.[1] At this time it "formed a beautiful promontory, jutting far out into the cove of waters, presenting on all sides extensive scenery, bounded only by the Jersey shore and the natural forest of the country."[2] It was also enclosed in a magnificent semi-circle of high hills, whose tops, covered with trees, reared their beautiful heads to heaven, forming one of the most splendid landscapes the world has ever produced.

On these rocks the Key of Kalmar and the Griffin landed their passengers and freight. They at once commenced the erection of a fort and trading house, which they named in honor of their young queen Fort Christina. They also changed the name of the creek to Christina Creek.[3] A small town, named Christinaham, or Christina Harbor, was also erected behind the fort. Lindstrom, an engineer who came out in 1652, left a plan of this town and fort.[4] It was butlt close to the point of rocks, its southern rampart being within a few feet of the creek.[5] On

[1] Ferris' Original Settlements. [2] Ibid.

[3] It was afterwards corrupted to Christiana its present name.

[4] See the plan in Campanius' Work.

[5] Various discoveries and relics have been made at different times in digging at the site of the fort. In 1745 a Spanish privateer threatened to land on the Delaware, and fears being entertained that they would attack Wilmington, attempts were made to place the old fort in repair. In digging the ground for that purpose, they found several

the eastern side of the fort, immediately under its walls, was a small cove or basin, called the "Harbor," in which their vessels might lay out of the current of the Christina, and without danger from the floating ice, on the breaking up of winter. "This basin is now filled up, and the cattle are browsing where their ships were once moored, but its original outline and form are yet distinctly visible, coinciding precisely with the representation made by Lindstorm," over two hundred years ago.[1] An Indian sachem, named Mattahoon, lived near the fort, from whom Minuit bought the land, for which he gave him a copper kettle, and some other small articles. He also bought of the same Indian as much land as was contained with "six trees." For this, the Indian afterwards stated, Minuit promised him half the tobacco which would grow upon it, which, however, the sachem said he never gave him.[2] This Mattahoon was undoubtedly the sachem Matta Horn who called a council in 1645, to consider whether the Indians should destroy the Swedes.[3]

The arrival of the Swedes was almost instantly known to the Dutch who inhabited Fort Nassau, as

pieces of money, with Queen Christina's stamp upon it. On the 31st of March, 1755, on taking up by chance some pieces of the walls, there were found many cannon balls, granadoes, and other similar things, which had been kept carefully concealed since the surrender of the fort by Rising. Five pieces of cannon (according to Acrelius) were kept mounted there previously, as at the treaty of Aux la Chapelle, in 1646, an English salute was fired from them, in honor of the governor, who was going to meet the Legislature at New Castle.

[1] Ferris' Original Settlement, p. 43.

[2] O'Calligan, vol 1.

[3] See ante p. 81.

on the 28th of April the assistant commissary of that fort wrote to New Amsterdam that Minuit had "sent his ship below the fort, and afterwards wanted to send her up again, but this we prevented." The commissary of Fort Nassau also sent Peter Mey to Minuit to see his license and commission, but this he refused to show. Jan Jansen, the clerk, was ordered to protest in proper form, in case Minuit did any thing to their disadvantage. As to whether Jan Jansen protested or not, history is silent. But the following formal protest was almost immediately made by Director Kieft,[1] from Fort Amsterdam:

[1] The character of Kieft stands out darkly in comparison with that of Peterson De Vries, who first attempted to settle this State. He thus describes the massacre of the Indians at Pavonia, now Jersey City, opposite New York, under the orders of Kieft:

"It was on the nights," he says, "of the 25th and 26th of February, 1643, that they executed these foul deeds. I remained that night at the governor's and took a seat in the kitchen near the fire. At midnight I heard loud shrieks. I went towards the parapets of the fort, and looked towards Pavonia. I saw nothing but the flash of the guns, and heard nothing more of the yells and clamor of the Indians, who were butchered during their sleep. About day the soldiers returned to the fort, having murdered eighty Indians. And this was the feat worthy of the heroes of old Rome, to massacre a parcel of Indians in their sleep—to take the children from the breasts of their mothers—to butcher them in the presence of their parents, and throw their mangled bodies into the fire or water. Other sucklings had been fastened (by their mothers) to little boards (according to the Indian manner of nursing very young infants) and in this position they were cut to pieces! Some were thrown into the river, and when the parents rushed in to save them, the soldiers prevented their landing, and let the parents and children drown together! Children of five or six years old were murdered, and some aged, decrepid men cut to pieces. Those who had escaped these horrors and found shelter in bushes and reeds, making in the morning their appearance to beg some food to warm themselves were killed in cold

"I William Kieft, Director-General of New Netherlands, residing on the Island of Manhattan, in New Amsterdam, under the sovereignty of their High Mightinesses the State General of the United Netherlands, and the privileged West India Company's department at Amsterdam, make known to the Hon. Peter Minuit, who calls himself commissioner in the service of her royal majesty of Sweden, that the whole South River, in New Netherlands, has been in our possession many years, and has been secured by us with forts above[1] and below,[2] and sealed with our blood, which has happened even during your direction of New Netherlands, and is well known to you. Whereas you now do make a beginning of a settlement between our forts, and are building there a fort, to our prejudice and disadvantage, what we shall never endure or tolerate, and which we are persuaded it never has been commanded by her royal majesty

blood, and thrown into the fire or water. Some came running to us in the country with their hands cut off. Some had their arms and legs cut off. Some who had their legs cut off were supporting their entrails with their arms. Others were mangled in other horrid ways, in fact too shocking to be conceived."

De Vries, in his remonstrance against the above massacre: "Consider, sir, what good will it do? We know that we lost our settlement at the Hoorn Creek (Lewestown) in 1630, by mere jangling with the Indians, when thirty-two of our men were murdered by the Indians, and now lastly at Staten Island, where my people were destroyed, occasioned by your petty contrivances in killing the Indians at Raritan, and mangling the brother of their chief for a mere bagatelle." De Vries, notwithstanding all his losses by the Indians, had a good opinion of them. He says of them, "they will do no harm if no harm is done to them."

[1] Fort Nassau.

[2] Fort Oplandt, at Lewes, where De Vries' men were massacred.

of Sweden, to build fortresses on our rivers, and along our shores, so is it that we, if you proceed with the building of forts, and cultivating the lands, and trading in furs, or engage further in any thing to our prejudice, protest against all expenses, damages, and losses, and will not be answerable for any mishaps, effusion of blood, troubles and disasters which your company might suffer in future, while we are resolved to defend our rights in all such manner as we shall deem proper. Done in the year 1638."[1]

No attention was paid to this protest by Minuit, who went quietly to work to finish Fort Christina, and was submitted to by the Dutch. Shortly afterwards the Swedes purchased all the lands from the Indians from Cape Henlopen to Santickan, or what is now known as the Falls of Trenton, and there fixed up stakes and marks. The original deeds for these lands, with the marks of the Indians, were sent to Sweden, and preserved in the archives of Stockholm, where they, as well as a map of the country, made by Magnus Kling, their surveyor, was seen by Israel Helm, and a copy of the map made and brought over by a clergyman who arrived here in 1697.[2] Part of this land, supposed to extend from Cape Henlopen to Mahon River, had been sold eight years before to Godyn.[3] It was on this land that De Vries' colony

[1] This document is dated Thursday, 6th of May, 1638, in Acrelius' History of New Sweden. At the records in Albany, N. Y., Monday, the 17th day of May. The Swedes used the old, the Dutch the new style; hence the difference of days. The reader must keep this in mind in regard to dates.

[2] Rudman's notes, in Clay, p. 17.

[3] See ante p. 127.

was massacred. The Indians thus sold to the Swedes what they had previously conveyed to the Dutch. We must not judge them harshly for this, for the Dutch had to all appearances abandoned the territory, and without doubt they were not aware they were committing any injustice or wrong in selling it over again. It was impossible for them to be acquainted with European rules in relation to real estate conveyances. To them the land was hunting ground, free to all. It will be seen hereafter they even denied to the Dutch the selling of the land to the Swedes, and sold it to the Dutch a second time. This leaves the matter a question of veracity between the Swedes and Indians.

As to who had the best right to the Delaware, the Dutch or the Swedes, we think that hardly any candid person will deny that, according to the European rule of right, the better claim lay with the former. According to the claim set up by the European maritime nations, the right of first discovery gave the discoverer the paramount right of occupancy and possession, and of purchasing from the Indians. According to this rule, the Dutch had a prior right to the Swedes, as they had discovered the Delaware in 1609, sailed up it as high as the Schuylkill, and built Fort Nassau, and afterwards Fort Oplandt, on the sight of or near Lewistown, in Sussex county, in this State. But before this the English had sailed past and discovered the shores—and it is said mapped them out—and three years previous to the Dutch discovery of the Delaware, had laid claim to the territory, and granted it to companies for settlement, who had settled portions of it under that grant. So that according to these rules, neither Dutch or Swedes had a right, but only the English. But according to natural right, the Swedes had a better claim than the Dutch, at least to Fort Christina, and the places they occupied, if not to the territory from Mahon River to the Trenton Falls. For the Indians were an independent people, lords of the land, had a perfect right to make sales to whoever they chose; and if they sold the Swedes the land, as the Swedes claimed they did, they had a better right than either the Dutch or English, who really had no right at all.

The probable reason of the Dutch submitting so quietly to what they considered the usurpations of the Swedes, was that the charter of the West India Company prohibited their declaring war or commencing hostilities either with a foreign State or the native Indians, without the consent of the States General of the United Netherlands; and in case a war should be waged against the company or its settlements, the States were only bound to furnish one half the means for equipping and manning a squadron for the occasion, and after it went into service the expense of maintaining the armaments were to be paid wholly by the company. Again, Sweden was then a great military power, and considered the champion of Protestantism in Europe, at the time when the feeling between it and Catholicism was most intense, and there was a struggle between them not only for dominion, but on the part of the former almost for existence.

Ferdinand II., Emperor of Germany, a cruel and despotic prince, had determined to extirpate the Protestant religion from every part of his extensive dominions. Gustavus Adolphus, supported by other European powers, determined to invade Germany. In the year 1630, he crossed the Baltic, and, after gaining several brilliant victories in a short time, took three hundred strong towns and fortresses in the German empire, and gained possession of the extensive tract of country extending from the borders of Hungary and Silesia to the banks of the Rhine, and from the Lakes of Constance to the Black Sea. This success having prevented the execution of Ferdinand's designs against the Protestants, we can imagine, caused a fraternal feeling between the Swedes and the Dutch, who but a few years before had separated from Spain, on account of their religion, the main cause of the Dutch revolt being, that whilst they held the Protestant, the Spaniards held the Catholic faith. Therefore, in addition to the blood and treasure which might be spent, were an attempt made to dispossess the Swedes by force of arms, there was the rupture of good feeling that would occur between two Protestant nations, a feeling which was much stronger then than now, when religious persecutions are unpopular, and wars between nations on account of their faith unknown. It was from these causes, and not from any want of courage, that the Dutch submitted to the usurpations of the

Swedes on the Delaware. Both, however, in the end succumbed to a stronger and more numerous race, who now inhabit the land for which they disputed, and whose rule, language, manners and customs may possibly hereafter extend over the globe.

CHAPTER XII.

FROM 1638 TO 1642.

Departure of Minuit for Europe—The return—Complaints of Dutch Company against residents for infringing on their trade—Employees of Company forbidden to trade—Slavery on the Delaware—Dutch Company complain of the heavy expense of Fort Nassau, and of injury to their trade by the Swedes—Dutch forbid powder and ball being sold to Indians—They forbid any vessel to sail in South River—Penalty—Swedes think of abandoning Delaware and going to New Amsterdam—Prevented by the arrival of the ship Fredenburg with succor—Arrival of Hollendare—The Dutch settlement under the Swedish rule—Their Charter—Jost de Bogart, their Governor—It is supposed they settled at St. Georges and Appoquinimink Hundreds—Complaint by the Dutch of injury to trade and outrage from the Swedes—English settle on the Delaware—Kieft's protest against them—Death of Peter Minuit, the first governor of Delaware—New Sweden the name of Delaware.

About three months after Minuit entered the Delaware, and probably directly after the Fort was finished, he sailed again for Europe, with the two vessels, leaving twenty-four men in the fort. But he returned soon afterwards.[1] There is, however, little left to tell us of the actions of the first governor of our State.

The Dutch West India Company made great complaints, on the 7th of June, of the frauds committed against them in the fur trade by parties resident in New Netherlands, whom they asserted "embezzled

[1] Hazard's Annals, p. 48.

and appropriated to themselves the largest and choicest assortment of furs, exchanging their worst skins for the best skins of the company." They complained grievously of the expenses of vessels and forts, and forbid any merchandise being sent, without the consent of the Company, to New Netherlands. They also forbid any person from trading without their consent, under penalty of loosing all their wages.

At this early period there appears to have been slavery on the Delaware. [1639] As one Coinclisse was "condemned, on the 3d of February, to serve the company with the blacks on South River for wounding a soldier at Fort Amsterdam. He was also to pay a fine to the fiscal, and damages to the wounded soldier."[1] On the 22d, a witness testifying in the case of Governor Van Twiller, (the governor of New Netherlands before Kieft,) who was charged with neglect and mismanagement of the company's affairs, said that "he had in his custody for Van Twiller, at Fort Hope and Nassau, twenty-four to thirty goats, and that *three negroes bought by the director* in 1636, were since employed in his private service."[2] Thus it will be seen that slavery was introduced on the Delaware as early as 1636, though probably not in this State, as the Dutch at that time had no settlement here. By another witness in the same case, we find that a large house was built at Fort Nassau, which was much decayed. The direc-

[1] Albany Records, vol. 2, p 10.
[2] Albany Records, vol. 1, p. 85.

tors in Holland severely condemned Van Twiller for these expenses. The mention of these matters are necessary as showing beyond a doubt the continued occupancy of this river by the Dutch. The directors also complain that Fort Nassau is a heavy burden to them as regards the garrison, provisions, and the vessel. They say, however, they cannot reduce it on account of the Swedes, who, being but five Dutch miles (twenty English) have done 30,000 florins worth of injury to their trade. They however consoled themselves by the thought the Swedes would soon have to break up if they received no succor.[1]

Thus early had the arrival of the Swedes in this State injured the Dutch trade on the Delaware. They had been very successful in their trade the first year after their arrival. They had exported *thirty thousand skins,* having managed, in their dealings with the Indians, to undersell the Dutch.

On the 31st of March, the Dutch made several rules in relation to the South (Delaware) River. Amongst them was one prohibiting the selling of powder and guns to the Indians, under penalty of death. Also one prohibiting any one sailing with boats or vessels on the South River without license, under the penalty of the confiscation of the vessel and cargo.

[1640] Notwithstanding the excellence of their trade, the Swedish settlers at fort Christina became discouraged, and in the spring had determined

[1] Holl. Doc., vol. 8, pp. 52, 53.

to abandon their settlement, and remove to New Amsterdam.[1] Every preparation had been made for their removal, and they were to have started the next day, when, to their great joy, a vessel arrived with succor from Sweden. This, of course, caused them to abandon their intention. She proved to be the Fredenburg, Captain Jacob Powelson. She was a Dutch vessel, and her crew and passengers were Hollanders, though sailing under the Swedish flag. She had left Holland in January, and arrived in the South River some time previous to the 1st of May. She was laden with men, cattle and everything necessary for the cultivation of the country. They brought with them a letter from the two Oxenstierns (the chancellor and his brother), directed to the commissary or other inhabitants of Fort Christina, in New Sweden, commanding that no obstacle should be placed in their way by the Swedish inhabitants, and that for their own advantage and safety, they should be on good terms with them. The same letter informed them that two more vessels would be sent out to them in the spring.[2] In this vessel it is said arrived Peter Hollendare, who succeeded Minuit as Governor of Delaware.

This settlement of Hollanders was made under a charter first given to Gothardt de Redden, William de Horst and —— Fenland, but it was afterwards

[1] Holl. Doc., vol. 8, pp. 53, 54.

[2] See Register of Penn. vol. 4, 179, where are published from documents furnished to the American Philosophical Society, by Jonathan Russel, whilst minister to Sweden.

assigned (for what reason, at the present day is not known), to Henry Hockhammer, &c., they, as the instrument states "having the intention of establishing a colony in New Sweden." As what now constitutes the State of Delaware was then entirely under the dominion of the Swedes; what Dutch there were being settled in Pennsylvania or New Jersey, we may consider New Sweden, (although its boundaries were undefined,) as being the appropriate name of Delaware at this time. The principal features of this charter were, that provision was given them under royal protection, to depart from Holland at their own expense, with two or three vessels, with "men, cattle and other necessaries to form a settlement, at least five German miles below Fort Christina, on both sides of South River, and take up as much land as they could place in actual cultivation in ten years." In addition to the fullest power over the land, if one part did not suit them they were allowed to choose another, with the consent of the Governor of the country. They were to pay to the Swedish crown as an acknowedgment of its authority, three florins for each family established in the territory. They had the right of exercising high and low justice, of founding cities, villages, and communities, "with a certain police, statutes and ordinances, to appoint magistrates and officers, to take the title and arms of their colony or province." "It being," the charter says, "understood that they and their descendants shall receive of us and our successors, that jurisdiction, and these royal rights, as an hereditary fief, and that they must conform themselves in this case, to all which concerns the ordinary justice of fiefs." Their statutes and ordinances were to be communicated to the governor for approbation and confirmation. They were allowed to exercise "besides the Augsburg confession," the "*pretended reformed religion*," but "in such a manner, that those who profess the one or the other religion, should live in peace, abstaining from every useless dispute, from all scandal and all abuse." They were "to be obliged to support at all times as many ministers and school-masters as the number of inhabitants shall seem to require," and "to choose for this purpose, persons who had at heart the conversion of the pagan inhabitants to christianity." They were allowed to establish all sorts of manufactories, and to "engage in all commerce, in and out of the country, with the coast of the West Indies and Africa, belonging to friendly powers, but only with vessels and yachts built in New Sweden, under promise of the government's assistance. Gottenberg was to be the depot of all merchandise transported from New Sweden to Europe. The merchants were not required to pass the sound, if they wished to go to some other part of the kingdom. They were not to enter foreign ports not dependant upon the crown of Sweden, except in cases of necessity, and they were then to proceed to Gottenberg, to show the reasons that compelled them to enter the foreign port, to pay duty on the merchandise they had sold elsewhere, and to equip their vessels anew. They were to be exempt for ten years, from all impost and duty (except the three florins on each family). After the ten years they were to pay five per cent. on all goods that should be imported or exported from New Sweden, and contribute to the pay of officers and support of fortresses. No one was allowed to take a servant from his or her master, before the term of service was ended, nor employ them without express permission of the master or governor, who was required to support the master in his rights. Whoever discovered mines, minerals, or precious stones, were to have the right of working them for ten years, with the consent of the governor,[1] after that to have the preference, upon the payment of an annual sum to be determined. Property was to be exempt from confiscation, and no fines, whatever be the offence, were to exceed one hundred florins of the empire, or forty rix dollars.[2] There were various other matters in the charter, but these appeared to be its principal features.

From the above it appears that a separate Dutch colony was formed, (within the present limits of this state), under the Swedish rule. Jost de Bogardt it is believed was appointed their governor, at a salary of 500 florins or 200 rix dollars per annum, with the promise of an increase of 100 florins to his salary, "if" (as his commission says,) "in future we have new proofs of his attachment, and of his zeal to promote our welfare, and that of our crown.[3] As to

[1] From this sentence it would seem that the consent required was that of the Dutch governor, appointed over these particular settlers, and not the Swedish governor.

[2] Swedish Documents in Register of Pennsylvania, vol. 4, p. 170.

[3] Reg. of Penn. Acrelius.

how Jost de Bogardt acted, or whether he got an increase of his salary, history is silent. This Dutch settlement it is supposed, was in the neighborhood of St. Georges and Appoquinimink hundreds in New Castle county. The instructions to Governor Prince, which it is supposed alluded to this colony, says, "that those of the Dutch nation who have gone to New Sweden, and are there established under Swedish protection, and under commandant Jost de Bogardt, the governor must show them all good will, seeing that they comply with the conditions stipulated, and also obey the orders signified to them this year; that being established too near to Fort Christina (report says only three miles distant), they must abandon it, and occupy a place more distant from the fort; but the governor may remove them or suffer them to remain, as he finds convenient.

The Dutch at this time made most grievous complaints to the directors in Holland, of the injury done to their trade, and outrages by the Swedes on the South River. Director Kieft, in a letter dated October 15, said, "we shall treat the Swedes with every politeness, although they commenced with many hostilities forcibly to build, attack our fort, trading, threatening to take our boats, &c."[1]

At this time, we have the first authentic information of the attempt of the English to settle on the Delaware. The different records do not exactly agree, but enough can be extracted from them to render it certain that Captain Turner, as the agent of several

[1] Holl. Doc., vol. 8, pp. 53, 54.

citizens of New Haven, embarked from there for the Delaware, and when he arrived, purchased several tracts on both sides of the river, for the purpose of settling it with English families from that place. On their way they called at New Amsterdam, and Director Kieft made a protest against their visit. But upon representation that they had "express direction not to meddle with anything that the Dutch and Swedes had a right too." He wrote to John Johnson (Jan Jansen in Dutch), the Dutch agent at Delaware to hold "good correspondence with them," which accordingly at first he did, and showed them how far the Dutch and Swedish title reached. The rest he told them was free for them to purchase, and offered them his assistance.[1] Trumbull in his history of Connecticut, says, "A large purchase, sufficient for a number of plantations was made by Captain Turner, agent for New Haven, on both sides the Delaware Bay and River. This purchase was made with a view to trade, and for the settlement of churches in gospel order and purity."[2]

[1641] Early in the year, the English made their attempt to settle on the lands purchased by Turner on the Delaware. A bark or catch was fitted out at New Haven, by Mr. Lamberton, under the command of Robert Cogswell. They stopped at Fort Amsterdam, when Director Kieft hearing of their intention delivered them the following protest:

[1] Records United Colonies in Hazard's Historical Collection, vol. 2, p. 213.

[2] Trumbull, vol. 1, p. 116.

"I, William Keift, Director General, &c., make known to you Robert Cogswell, and your associates not to build nor plant on the South River, lying within the limits of New Netherlands, nor on the lands extending along there, as lawfully belonging to us, by our possessing the same long years ago, before it was frequented by any Christians, as appears by our forts which we have thereon; and also the mouth of the rivers sealed with our blood, and the soil itself, most of which has been purchased and paid for by us, unless you will settle under the States and the noble West India Company, and swear allegiance, and become subject to them, as the other inhabitants have done. Failing whereof, we protest against all damages and losses which may accrue therefrom, and desire to be holden innocent thereof."[1]

The English however, assured the Governor, that it was not their intention to interfere with any settlement already made, and if none could be found free from claims they would return. They accordingly proceeded, to South River, and commenced erecting trading houses or Varkens, or Farkenskill, near the present town of Salem, N. J., and it is supposed also on the Schuylkill. This settlement consisted of sixty persons,[2] comprising twenty families.[3]

About this time died Peter Minuit, the first Governor of New Sweden. He is usually placed in our annals, as the first Governor of Delaware, although

[1] Holl. Doc., vol. 9, p. 205.

[2] Instruction to Swedish Governor, Reg. Penn. vol. 4, p. 219.

[3] Winthrop.

that claim rightly belongs to Peter Hosset. To his exertions our State owes its first permanent settlement. He was buried it is supposed, in what is now the graveyard of the Old Swedes Church, in the City of Wilmington. There his bones rest, in company with those of Reorus Torkillus, the first clergyman. Thus, the first chief magistrate and the first minister of Delaware, sleep side by side, with no stone or insignia to mark their resting place.

It is supposed, that the Old Swedes Church is built on the site of the old graveyard, used by the first Swedish settlers to inter their dead. This is inferred from the following circumstances: Fort Christina served the purpose both of a church and fortification, as in it the Swedes held their public worship. In the rear stood the village of Christinaham, the ancestor of the modern Wilmington. On the west side of Christinaham being near to the village, a few hundred yards from it, was the present cemetery of the Old Swedes Church. For many years the great body of the settlers clustered in their habitations round the fort. All the rest of the country at that time was wild and uncultivated. It is not to be supposed, that a people so religious as the Swedes, would be without a burial place. No other graveyard was ever known at that time to be laid out in the vicinity. The contract for the erection of the Old Swedes Church said "that it was to be built in and upon the churchyard at Christeen." This proves it was a burial place before building the church. Therefore, judging from these circumstances the probabilities are, that the churchyard of the Old Swedes Church, was the first cemetery in Delaware, and that there was buried Peter Minuit and Reorus Torkillus.

Minuit appears to have been an energetic, just man. Acrelius says, "he did great service to the first Swedish colony, during three years he protected this small fort (Christina), which (in his time) the Dutch never attempted. Vanderdonk, in his description of New Netherlands, quotes of a letter of Minuit's whilst Governor at New Amsterdam, which shows his concern for the agricultural interests of the colony. He tells us "that Mr. Minuit writes that he has sown canary seed, and that it grew and yielded well," but he adds, "that the country is new, and in a state of beginning; and that the time of the cultivators should not be spent in such experiments, but to the raising of the necessaries of life; of which, God be praised, there is plenty and to spare for a reasonable price. And we begin to supply provisions and drink in common with our Virginia neighbors, to the West Indies, and to the Carribbee Islands, which we expect will increase from year to year, and in time become a fine trade, in connection with our Netherlands and Brazil commerce." And again, "Commander Minuit testifies that cremmin seed, canary seed, and the like have been tried, and succeed well, but are not sought after."[1] But little however is known of him, save a few detached items which we get from the correspondence of the period.

[1] Vanderdonk's N. Y. His. Coll., pp. 156–160.

CHAPTER XIII.

FROM 1642 TO 1643.

Peter Hollendare succeeds Minuit as Governor of New Sweden, (Delaware)—He returns to Sweden—Made Governor of the Naval Arsenal—English expelled from Schuylkill by the Dutch—Col. Jno. Printz appointed Governor of New Sweden—His instructions in relation to the government of the country—In relation to the English—They were to be removed if possible—The Dutch right to the South River denied—They are not to be allowed to stop the Swedes from trading above Fort Nassau—If they use force, they are to be repelled by force—Savages to be treated with humanity—To be instructed in the Christian religion—Goods to be sold them lower than by the Dutch and English—Fortifications to be erected to shut up or command the South River—Offences to be punished—Death inflicted in some cases—Fisheries to be established—Salt and oil made—Timber exported—Religion to be according to the Augsburg Confession—Dutch allowed to hold the reformed faith—Expenses of government of New Sweden.

[1642] AFTER the death of Minuit, Peter Hollendare became Governor of the Swedes on the Delaware. He was an officer of the Swedish army. Of his character as a governor nothing is known. He held his office for about one year and a half. He then returned to Sweden, and was made commander of the Naval Arsenal at Stockholm. But little is known of his acts.[1]

The Dutch at New Amsterdam, now determined to

[1] Acrelius, p. 410; O'Calligan, vol. 1, p. 366; Clay, p. 17.

expel the English from the Schuylkill, the settlement at Varken's kill or Salem, they appear at this time not to have interfered with. The Director General and Council at their meeting passed a resolution which says, that "having received unquestionable information, that some English had the audacity to land at South River, opposite to our Fort Nassau, where they made a beginning of settling on the Schuylkill, without any commission of a potentate, which is an affair of ominous consequence, disrespectful to their High Mightinesses, and injurious to the interests of the West India Company, as by it their commerce on the South River might be eventually ruined:" Resolved, "that it is our duty to drive these English from thence, in the best manner possible." Accordingly on the 22d of April, they issued instructions to their commissary or governor on the South River, Jan Jansen Van Ilpendam (called by the English John Johnson), that as soon as the yacht Real and St. Martin should arrive at the South River, he was to embark in either of the two yachts, (or, if he thought best in both of them,) with such a body of men as he could collect together, and proceed towards the Schuylkill, disembark there directly, and require from the English to show him by what authority they acted, and how they dared to make such an encroachment upon "the Dutch rights and privileges, territory, and commerce." If they could show "no authority or royal commission to settle within," the Dutch "limit," or "an authentic copy of such a commission," then then they were to be compelled "to depart directly in peace, to prevent effusion of blood." If they would "not listen or submit," then "their persons" were to be secured and brought to New Amsterdam. He was instructed to be on his guard, to remain master, not to be surprised, and to "maintain the reputation of the High Mightinesses and the noble Directors of the West India Company." If the English left the spot, or made their escape then he was to destroy their improvements, and level them "on the spot." Whilst he was to take care that the English were not injured in "personal property," but that "in their presence" an accurate inventory should be made of the whole.[1]

Jan Jansen appears to have carried out his instructions, and have expelled the English from the Schuylkill, for the Dutch chronicles are silent as regards the attack on the English settlers. The complaints of the latter, however, are made known in the English records at New Haven. They say that notwithstanding the purchases of the English on both sides of the river, to which they affirm neither the Dutch or Swedes had any just title; Governor Kieft, without protest or warning sent armed men and by force, in a hostile manner burnt their trading house, seized and for some time detained the goods in it, not suffering their servants so much as to take an inventory of them. He also seized their boat, and for a while kept their men prisoners, for which treatment they could not up to 1650, get any satisfaction. They also assert that they "attempted to seize Mr. Lamberton's

[1] Albany Records, vol. 2, pp. 162, 164, 165.

vessel, or drive him out of the river," but being on his guard, he at that time maintained the right and honor of the English. As he was returning from Delaware, the Dutch Governor at Manhattan compelled Mr. Lamberton, who was the agent at New Haven, "by threatenings and force, to give an account of what beaver he had traded for within the English limits at Delaware, and pay recognitions for them, against which a protest sent from New Haven was of no avail." The damages done to the English at Delaware was estimated at £100[1] sterling.[2]

At this time a great sickness and mortality prevailed among the settlers on the Delaware. It affected both Swedes and English.[3]

On the 16th of August, John Printz, a Lieutenant-Colonel in the Swedish army, was appointed Governor of New Sweden. He was the third Governor of Delaware. His instructions were dated Stockholm, the 15th of August, 1642, or one day before his appointment. His commission was as follows: "Our faithful subjects having commenced visiting the West Indies, and having purchased in form, and already occupied a considerable part of that country, which they have named New Sweden, in consequence—as their laudable project, the navigation which they have undertaken, and the cultivation which they are disposed to make, cannot but increase and facilitate commerce—to give them more vigor and extent, not only

[1] About $500.

[2] Hazard's Historical Collection, vol. 2, pp. 162, 210.

[3] Winthrop, vol. 2, p. 76.

have we approved their design, and taken the country and its inhabitants under our royal protection, but again to favor and strengthen the work which they have commenced, we have given to the country and inhabitants, our subjects, a Governor, and have named as we do here, by virtue of his letter patent, our very faithful subject, the above named Lieutenant of Cavalry, John Printz for Governor of New Sweden. He engages to administer and govern said country, and to defend its inhabitants against all violence and foreign attachment, and to preserve above all, that country in safe and faithful hands. He must preserve amity, good neighborhood and correspondence with foreigners, with those who depend on his government and the natives of the country; render justice without distinction, so that there shall be injury to no one. If any person behave himself grossly, he must punish him in a convenient manner; and as regards the cultivation of the country, he must in a liberal manner regulate and continue it, so that the inhabitants may derive from it their honest support, and even that commerce may receive from it a sensible increase. As to himself, he will so conduct his government, as to be willing and able faithfully to answer for it before God, before us, and every brave Swede, regulating himself by the instructions given to him." The inhabitants are required to acknowledge and obey him as Governor.[1]

His instructions, which he received a day earlier,

[1] MS. Doc. American Philosophical Society, Reg. of Penn. vol. 4, p. 200.

were as follows: After reciting the advantages expected to result from "the conquering and purchasing the territory of New Sweden, and the extension of commerce thereby," and that "for this laudable end, two vessels named the 'Stork' and the 'Renown,' have been furnished," he was instructed to go to Gottenberg, to embark from there, to consult with the captain and council as to the manner of the voyage, the course he should take, &c. When he arrived in New Sweden, he was to take care, according to the contract with the savages, "that the *frontiers of the country extend from the borders of the sea to Cape Henlopen, in returning southwest towards Godyn's Bay, and thence towards the Great River, as far as Minquas kill, where is constructed Fort Christina, and from thence, again towards South River,* and the whole to a place which the savages called Santickan, which is at the same time the place where are the limits of New Sweden. This district or extent of country may be in length about thirty German miles; as to width in the interior, it has been stipulated in the contracts, that the subjects of her majesty and company may take as much of the country as they wish." The boundaries claimed by New Sweden, were thus, from Cape Henlopen unto the city of Trenton, N. J., comprising the whole of Delaware and part of Pennsylvania, including the ground on which Philadelphia now stands.

As regards the English settled on "Ferkin's kill," as the agents of the Company, as her majesty's subjects had bought from the Indian owners, the whole district from Cape May to Raccoon Creek, with a view to unite these English with the Swedes; the governor was faithfully to perform the contract. His instructions also say, that "she (the Queen), suggests that these people are disposed to submit as a free people to a sovereign who can protect and defend them, and advises a conciliatory course," as yet, as her majesty judges it will be better and more advantageous for Sweden, for the crown, and for those interested, to be able to disembarrass themselves of it honestly, she leaves it to the governor's discretion, to endeavor to find this point, and for it to work understand as much as possible, with good manners and with success.

As regards the Dutch. The instructions after reciting their claims to South River, and stating that they (the Dutch), lay no claim to the "western part of which the Swedes are in possession." After also asserting that "they undoubtedly wish to appropriate to themselves the lands possessed by the English, and certainly all the eastern part of the great South River," and have endeavored to prevent the Swedes from ascending above Fort Nassau, he was instructed "to comport himself towards them with mildness and moderation," as the Swedes "only sought to open a free communication for commerce," and had "bought from the natives that which they possessed and cultivated." If, however, contrary to all hope, the Dutch should "show any hostile intentions, the instructions say, "it would be very very proper to be on your guard, and repel force by force," "at so great a distance the government leaves it much to the Governor's discretion." If the Dutch did not "trouble him, he was to maintain amity and good neighborhood with them, also with those who inhabited Manhattan or New Amsterdam, and likewise with the English who inhabited Virginia, especially as the latter had began "to procure for the Swedes all sorts of necessary provisions, and at reasonable prices, both for cattle and grain."

As to the Dutch who had gone to New Sweden, and were there established under Swedish protection, under Commandant Jost de Bogardt; the government must show them all good-will, seeing that they comply with the conditions stipulated, and also obey the orders signified to them this year, that being established too near Fort Christina (report says only three miles distant), they must abandon it, and occupy a place more distant from the Fort; but the governor may remove them or suffer them to remain as he finds expedient.

He was instructed to treat "the savages with humanity and mildness and see that neither violence nor injustice was done them," but "must labor to instruct them in the Christian religion, and the divine service, and civilize them." He was to "bring them to believe that the Swedes have not come there to do them injustice, but rather to procure them what they need, in order to live reciprocally in common, and sell and exchange provisions." He was instructed "to sell to them at lower prices than the Dutch at Fort Nassau, or the English; so as by this means to disengage them from these people, and accustom them more to the Swedes.

He was also instructed to choose his residence where convenient. To erect fortifications at Cape Henlopen, James' Island,[1] or any other favorable position. The fortress erected was to be able "to shut up the South River," or it must be commanded by it. If, however, he could protect himself with Fort Christina, he was instructed to turn his attention especially to agriculture, sowing enough grain for their support. He was to attend to the culture of tobacco, to raise sheep and cattle, to improve the breed by procuring from the English or others, and increasing their number. He was to have "commissaries to inspect the trade with the Indians, and prevent others from trading with them." The manufacture of salt was recommended, also examinations for "metals or minerals." He was to reflect on what was to be done with the superfluous wood, whether it could not be used for ballast, whether oil could not be procured from the nut trees by pressure. Whether fisheries, especially for whales, could not be established as they were especially numerous in Godyn's (Delaware) Bay, whether silk and silk worms might not be produced to advantage.

Justice was to be done in the name of "her majesty." "Detailed and perfect instruction (the instructions say) cannot be given, therefore it is left to the instruction of the Governor, according to circum-

[1] A part of Camden was formerly an island, and called James Island. See Mickle, p. 85.

stances." Controversies were to be decided by the "laws and customs of Sweden." He had power to punish for offences, "great offenders by imprisonment, and even with death, according to the crime, after legal forms and sufficient examination by the most noted persons, such as the most prudent assessors of justice that he can find and consult in the country," but "before all," he was "to labor and watch that he renders in all things to ALMIGHTY GOD, the true worship which is his due." Divine service was to be "performed according to the true Confession of Augsberg, the Council of Upsal, and the ceremonies of the Swedish Church." The Dutch residing in Swedish territory, were not to be disturbed in the exercise of the reformed religion.

Governor Printz was appointed for three years, after which he was allowed to return, leaving a successor or viceroy in his place, or he might be reappointed. He was allowed 1200 silver dollars per annum, to commence January 1, 1643.[1]

The Swedish government at the same time passed an ordinance appropriating 2,619 rix dollars to be collected each year, from the excises upon tobacco for the expenses of New Sweden. The following were the expenses of the military department, viz.: governor, 1200 silver, or 800 rix dollars, half silver and half excise; lieutenant governor, 16 rix dollars per month; one sergeant-major, 10 rix dollars; one corporal, 6 rix dollars; one gunner, 8 rix dollars; one trumpeter, 6 rix dollars; one drummer, 5 rix dollars;

[1] MS. American Philosophical Society, Reg. of Penn., p. 219.

twenty-four soldiers, at 4 rix dollars; one paymaster, 10 rix dollars; one secretary, 8 rix dollars; one barber,[1] 10 rix dollars; one provost, 6 rix dollars, and one —— 4 rix dollars; making 185 rix dollars per month, or 3020 rix dollars per annum.

[1] Surgeons were then called barbers.

CHAPTER XIV.

FROM 1642 TO 1643.

Governor Printz sails from Stockholm, for New Sweden, in the ships Renown and Stork—Arrival in the Delaware—Storm in the Bay—The Renown runs aground—Arrival at Fort Christina—Rev. Jno. Campanius—He builds Fort Gottenburg—A mansion and church on Tinicum Island—He builds Fort Elfsberg—De Vries' compelled to strike his flag at Fort Elfsburg—Great size of Printz—Drunkenness of Swedes—Printz expels the English from Salem Creek—Takes Lamberton prisoner—Persuades Lamberton's men to accuse Lamberton of exciting Indians against Swedes—Lamberton's men refuse to do so—They are placed in irons—Lamberton pays beaver to get his liberty—Printz abuses the English—Dutch assist the Swedes in expelling the English—Lamberton complains to the Court at New Haven—Governor Winthrop requested to write to Swedes and Dutch—Vagabonds and malfactors sent to New Sweden—They are prevented from landing—Mortality among them on their return—Erection of Fort Schuylkill—Fort Kingsessing—Fort Korsholm by the Swedes at Passyunk—They stop the trade of the Dutch on the Schuylkill.

On the 16th of August, Governor Printz sailed from Stockholm, for New Sweden, in the ship Renown,[1] accompanied by the Stork. They took a southerly route, sailing by the Portugese and Barbary coast, passing the Canary Islands, arriving at Antigua on the 20th of December. There they spent their Christmas holidays, and were well entertained

[1] Campanius calls it the Fame. We have preferred to go by the instructions to Printz, and call it the Renown. Other writers speak of the two vessels as the Fame, and the Charitas.

[1643] at the Governor's house.[1] On the 3d of January they left Antigua, and sailing by way of St. Christopher, St. Martin's and other West India Islands, on the 24th of the same month found bottom, and on the 25th saw land on their left. On the 26th they were in the bay off the Whorekill (Sussex county), and on that and the 27th experienced a severe storm, accompanied with snow, when the Renown was run aground and lost three large anchors, a spritsail, and the mainmast, and experienced other damages. On the 15th of February they arrived at Fort Christina. They were five months or 150 days on their passage, from the time they left Stockholm, until they arrived at Fort Christina. With this expedition came a clergyman named John Campanius Holm, more generally called Campanius; rendered celebrated from being the first to translate Luther's catechism into the Indian language. Also from keeping a journal of his visit to New Sweden, from which his grandson Thomas Campanius Holm, wrote his celebrated "Description of the Province of New Sweden."[2]

[1] Campanius.

[2] John Campanius Holm was born at Stockholm, on the 15th of August, 1601. His father was Jonas Peter, clerk of the congregation of St. Clara. He went through his studies with great reputation, and was for a long time preceptor in the Orphan's House, at Stockholm. On the 3d of February, 1642, he was called by the government to accompany Governor Printz to America, where he remained six years pastor of the congregation there. On his return home, he was made first preacher of the Admiralty, and afterwards was pastor of Frost Hultz and Herenwys Uplandt, where he translated Luther's catechism, with other things, into the American Virginia (Indian) language, a work which he had begun in America, and which he

Governor Printz, soon after he landed, agreeable to his instructions, selected a site for his residence, and commenced the erection of fortifications to command the river. At a place called by the Indians Tenacong (now called Tinicum), a short distance above where Chester now stands, lying between Darby and Crum Creeks, he found a beautiful piece of land, with a high bold shore. It is now the Lazaretto, used to quarantine vessels bound for Philadelphia with infectious diseases on board. It was then, as now, an island, having the Delaware on the east, Darby Creek on the south and west, and on the north, a sound or branch passing across a morass, and connecting Darby Creek with the Delaware near Fort Mifflin.[1] Here Printz built a fort which he named Fort Gottenberg, also a mansion. The fort was constructed by laying very heavy hemlock logs, the one on the other, and was "pretty strong."[2] The mansion for himself and his family, was "very handsome"; there was likewise a fine orchard, a pleasure house, and other conveniences. He called it "Printz Hall." On this island, the principal inhabitants (afterwards) had their dwellings and plantations. A church was also erected there, which, on the 4th of September, 1646, Dr. John Campanius consecrated for divine service, and also its burying place.[3]

here perfected. He died on the 17th of September, 1683, at the age of 82, and was buried in the church of Frost Hultz, where a handsome monument was erected in the choir to his memory.—*Campanius.*

[1] Ferris, p. 62.

[2] Hudde's Report, N. J. His. Society Mem., p. 420.

[3] Campanius, p. 79. Ferris says, this hall stood more than 120

This fort controlled the access of the Dutch to Fort Nassau. The same year he erected another fort at Varkenkill, afterwards called by the English, Salem Creek. It was called Elfsberg or Wootsessung,[1] afterwards known by the name of Elsinburg or Elsinboro. It was erected at the south side of the creek at its junction with the Delaware.[2] Hudde says, "It was usually garrisoned by twelve men, commanded by a lieutenant. It had eight iron and brass guns, and one potshoof.[3]

The main object of this fort was to visit the Dutch vessels, and oblige them to lower their colors as they sailed up the Delaware, which greatly "affronted them."[4] Peterson De Vries, the energetic projector of the unfortunate colony at Lewestown, was fired at, as he sailed up the Delaware, in the month of October of this year, and was ordered to strike his flag. He says the fort was commanded by Captain Printz, "who weighed upwards of *four hundred* pounds, and drank three drinks at every meal." He describes the Swedes "as not very sober, as they bought from the captain of the vessel, a good quantity of wine and sweetmeats, and that neither here nor in Virginia, was intoxication punished by whipping.[5]

Either shortly afterwards or previous to building this fort, Printz succeeded in expelling the English,

years, and was at last burnt down by accident, since the commencement of the present century, p. 70.

[1] Campanius. [2] Ferris.

[3] Hudde, p. 429. [4] Acrelius, p. 412.

[5] De Vries, p. 273. Hazard thinks this was not Governor Printz but a relative. Hazard's Annals, p. 73.

who were settled on Varkenkill, under Lamberton. He attacked them and burnt down their trading house, and by surreptitious means, succeeded in making Lamberton a prisoner.[1] Lamberton was in his pinnace named the "Cock," at anchor about three miles above Fort Elfsberg, when a letter was brought by two Swedes from Printz, ("Tim, the barber, and Godfrey the merchant's man,") stating that the Indians had that day stolen a gold chain from his wife, and that those Indians were about trading with Lamberton, and that he desired his good offices to get it back. He also desired Lamberton "to stay on board until the next morning," affirming, that "he would know the Indian that stole it by a mark that he had on his face." No Indians however, came on board. Lamberton afterwards calling at the Swedish fort, where, it is supposed, he went in obedience to a request from a second letter from Printz, was arrested, in company with Jno. Woollen, his Indian interpreter, and John Thickpenny, and placed in prison. Woollen was put in irons. Printz himself fastening them on his legs. It is asserted that Printz' wife, and Timothy the barber (surgeon), endeavored to get Woollen intoxicated, by giving him a quantity of wine and beer to drink, and that immediately after drinking the liquors, he was conveyed to Printz, who, "with professions of a great deal of love to him, making many large promises to do him good," endeavored to get him to say, "that George Lamberton had hired the Indians to cut off the Swedes." Woollen denied that Lamberton had

[1] Rudman, Swedish Records at Wicaco.

any such intention. The governor then "drunk to him again," and said, "he would make him a man, give him a plantation, and build him a house, and that he should not want for gold and silver," provided he made the accusation against Lamberton. But Woollen still refusing to accuse Lamberton, the governor was much enraged, and stamped with his feet, and calling for irons, "he put them upon Woollen with his own hands, and sent him down to prison." It is also asserted, that the watchmaster and Gregory, the merchant's man, endeavored to get John Thickpenny to accuse Lamberton of plotting with the Indians to cut off the Swedes.[1] But Thickpenny refused to make any such accusation.

Lamberton before he regained his liberty, had to pay a "weight of beaver" to Printz. Printz also expelled all the English that would not take the oath of allegiance to the crown of Sweden. He also railed at the English in a very intemperate manner. He cursed, swore, and reviled at them, calling them English renegades. The Dutch assisted the Swedes in the expulsion of the English from Varcken's (or Cohansey) Creek. Complaint of these outrages were made by Mr. Lamberton, to the court of New Haven, which met on the 2d of August. They were substantiated by the oath of John Thickpenny, one of Lamberton's sailors. The Court requested the President Governor Winthrop, to write to both the Dutch and Swedish governors, expressing particulars, and requiring satisfac-

[1] Deposition of John Thickpenny, New Haven Colonial Records, vol. 1, pp. 97–99.

tion; and professing, that as they would not wrong others so, they may not desert their confederate in a just cause.[1] A commission was also given to Mr. Lamberton, to go and treat with the Swedish government, about those injuries and damages, and to agree with him about settling their trade and plantation.

On the 7th of September, Reorus Torkillus, the first preacher in this State, or on the Delaware, died. He was born in West Gothland, in the year 1608. After going through his studies, he was made professor of a college at Gottenberg, and was afterwards chaplain to the superintendent, Andrew Printz, who was probably a relation to Governor John Printz. He fell sick on the 23d of February.[2] He married one of the residents of New Sweden (Delaware), by whom he had one child, whose descendants may possibly still be living amongst us, under an anglicized name. His death and burial have been before mentioned.

About this time, a number of robbers and malefactors were sent from Sweden, to settle on the Delaware. Campanius speaking of them, says: "The generality of people who went or were sent over from Sweden to America, were of two kinds. The principal of them consisted of the company's servants, who were employed by them in various capacities; the others were those who went over to that country to better their fortunes; they enjoyed several privileges; they were at liberty to build and settle themselves

[1] Hazard's His. Coll., vol. 2, p. 11.
[2] Campanius.

where they thought proper, and to return home when they pleased. By way of distinction, they were called *freemen*. There was a third class, consisting of vagabonds and malefactors; these were to remain in slavery, and were employed in digging the earth, throwing up trenches, and erecting walls and other fortifications. The others had no intercourse with them, but a particular spot was appointed for them to reside upon."

"In the beginning of Governor Printz's administration, there came a great number of those criminals, who were sent over from Sweden. When the European inhabitants perceived it, they would not suffer them to set their foot on shore, but they were all obliged to return, so that a great many of them perished on the voyage. It was after this forbidden, under a penalty, to send any more criminals to America, lest Almighty God should let his vengeance fall on the ships, and goods, and the virtuous people that were on board; it was said, that there was no scarcity of good and honest people to settle that country; but such a great number of them had gone thither (as engineer Lindstrom says), that on his departure from hence, more than a hundred families of good and honest men, with their wives and children, were obliged to remain behind, as the ship had taken as many on board as she could hold, and yet these honest people had sold all their property, and converted it into money, not imagining that they could be so disappointed."[1]

[1] Campanius, pp. 73, 74. He says, "This was related to me amongst other things, by an old trustworthy man, named Nils Matson Utter, who, after his return home, served in his majesty's life guards.

On the 2d of November, John Papegoy, who had previously been to New Sweden, received a letter of introduction to Governor Printz, from the Swedish government. The letter recommended his employment, protection, and advancement. He afterwards marrried Printz's daughter, and succeeded Printz as governor of New Sweden, until the arrival of John Claudi Rising. On the 6th of November of the same year, Queen Christina granted New Gottenberg or Tinicum Island to Printz.

In pursuance of his plan to fortify all important points on the Delaware, and "shut up the river," Printz erected a fortification on the Schuylkill. This river was so named by the Dutch. In their language it means "*hidden creek*, or Sculk Creek," from the retired and hidden situation of its mouth. This fort was built on an island in that river, within gunshot of its mouth. Ferris says: "At that time, all the great meadows extending from the high point of land at Bartram's Botanic Garden, (this garden does not now exist,) in a southerly course to the Delaware, were under water. When the tide was at its highest point, vessels drawing four or five feet water, could sail from Fort Gottenberg, or Tinicum Island, across the meadows to the mouth of the Schuylkill; which, at that period, opened just below the said garden, the south point of which was one of its capes. Just above the elevated point, on which stands Bartram's old mansion house, and through which, by a deep cut, walled on both sides, the Philadelphia, Wilmington and Baltimore Railroad passes; there is, on the present margin of the Schuylkill, a cluster of rocks, considerably elevated above the water, and partly covered with earth and forest trees. Between these rocks and what was once the shore, close by the railroad, there is a piece of meadow land more than two hundred yards wide; which, in Governor Printz's time was under water, and constituted part of the river Schuylkill. That cluster of rocks and the earth connected with them, formed the island on which Printz built the fort as aforesaid.[1] This fort was named Fort Manayunk, or Schuylkill. It was a handsome little fort, built of logs, filled up with sand and stones, and surrounded with palisades, cut very sharp at the top. It was mounted with great guns."[2] He also built a fort, or "strong house" at "Chinsessing," (Kingsessing.) This was called the "New Fort." "It was not properly a fort, but substantial log houses, built of good strong, hard, hickory, two stories high, which was sufficient to secure the people from the Indians." "In this settlement there lived five free men, who cultivated the land and did very well."[3] This fort was situated a little below the former fort. He also built a fort named "Korsholm," at Passayunk, in the same neighborhood. The tract of land on which this fort was built, was granted by the Swedish crown to lieutenant Swen Schute, who afterwards surrendered Fort Trinity, or rather Fort Cassimer to the Dutch. He was its commandant. "After Governor Printz's departure for Sweden, it was aban-

[1] Ferris' Original Settlements on the Delaware, pp. 70, 71.

[2] Campanius, p. 80.

[3] Ibid.

doned by the Swedes, and afterwards burnt and destroyed by the Indians."[1] Printz also constructed about half a mile in the woods, at "Karakung," otherwise called the Water Mill Stream, a "fine mill, which ground both coarse and fine flour." This was the first mill erected on the Delaware. "There was no fort near it, but only a strong dwelling house, built of hickory, and inhabited by freemen. This mill, Ferris supposes, was on Cobb's Creek which flows into Darby Creek. The site, it is supposed was on some rocks, just above the bridge where the Philadelphia road crosses that stream." The Dutch company's carpenter assisted the Swedes in the erection of one of these forts. One of the trading houses of the Swedes, was also built right before the Dutch Company's fort of Beversreede, not being a rod from the gate.[2]

From the above, we should judge the Dutch Company's employes were unfaithful to their interests. The conduct of the Swedish governor was also extremely insolent and tyrannical.

The erection of these forts, enabled the Swedes effectually to control the trade of the Schuylkill, the only remaining avenue for them to trade with the Minquas, without which trade, says the Dutch commissioner, Hudde, (who then had command of the Dutch fort on the Delaware,) this (the Delaware) river is of little value.[3]

[1] Campanius.

[2] Holl. Doc., p. 32, 50.

[3] Hudde's Report, p. 429.

CHAPTER XV.

FROM 1643 TO 1648.

Winthrop writes to Printz, complaining of the treatment of the English settlers—Printz denies the bad treatment—Expresses good feeling—Mr. Eaton asserts that English can settle on Delaware, on securing new commission—Expedition from Boston to Delaware to discover Lake Lynconnia—Jealousy of Dutch and Swedes—Drunkenness of the English captain—Swedish vessel from Delaware compelled to pay duty in Holland—Birth of William Penn—Queen Christina assumes the government of Sweden—Capture of an English vessel and murder of her crew—Removal of the Dutch governor, Jan Jansen—Appointment of Andreas Hudde—Destruction of Fort Gottenberg—Dutch vessel driven from the Schuylkill—Protest of Hudde—Dutch endeavor to ascend the Delaware above the falls—Are prevented by Indians through machinations of the Swedes—Grant of land opposite Reedy Island in Delaware—President Eaton of New Haven complains to Kieft of outrages to English on the Delaware—Dutch purchase land from Indians on site of Philadelphia—Erect Dutch arms upon it—They are pulled down by the Swedes, who protest against the purchase—Hudde's counter protest—Insult to the messenger—Bad treatment of Dutch by Swedes—George Lambert drowned—Stuyvesant appointed Governor of New Netherlands—He writes a complimentary letter to the Governor of Massachusetts and New Haven—Claims of the Dutch—Quarrels between Swedes and Dutch—Swedes accused of inciting savages against the Dutch—Vulgar language of Printz to Hudde—He takes powder and shot from a Dutch vessel—Vessels arrive from Sweden—Swedish tobacco ordinance—Swedish vessel refuses to show her colors at Fort Nassau—Swedes build on Schuylkill—Dutch receive grant of land from Indian sachems—Sachems tell Swedes they have no right to the land, but the Dutch have—Dutch attempt to build—Are driven off by the Swedes—Commissioners sent from New Amsterdam—Their impolite treatment by Printz—They protest against the Swedish outrages on the Schuylkill—Further outrages by Swedes—They drive away Dutch citizens, and threaten them with beating—Swedes build a house in front of Dutch fort of Beversreede—Murder of Swedes by Indians.

AGREEABLE to the instructions of the court, held at New Haven, Governor Winthrop wrote to Printz in relation to his treatment of Mr. Lamberton. [1644] Printz wrote back, denying the whole matter, "*using at the same time large expressions*" of his "respect" to the English, and particularly to the "*New Haven colony*." He also sent copies, on oath, of the "examination taken in the case, with a copy of all the proceedings between them" and the English who had settled on the Delaware from New Haven. These letters[1] were laid before the Court of the United Colonies of New England, which met at Boston on the 7th of March. It was also stated by Mr. Eaton that Printz requested to be shown a copy of the New England patent, and that he told the agent of the Commissioners of the United Colonies that upon a new commission from them he would allow the English to go on with their plantations on the Delaware Bay and River. This commission was granted.[2]

An expedition was sent from Boston to the Delaware to discover the great lake Lynconnia, which it was supposed laid northwest of the New England patent, and could be reached by sailing up the Delaware River. It was supposed that the great beaver trade "from the eastern and southern ports" came

[1] No copies of these letters are now in existence.

[2] Winthrop, vol. 2, p. 257.

from thence. They intended to sail in their pinnace, which was well manned and supplied with provisions and "trading stuff," "as high up as they could go, and then some of the company under the conduct of Mr. William Aspinwall, a good artist, and one who had been in these parts, to pass, by small skiffs or canoes, up the river as far as they could." They took with them letters both to Printz and Jan Jansen, the Dutch governor. The Dutch allowed them to go up, telling them, however, they would have to protest against them for their own interests. But the Swedes brought them too by a shot from one of their forts. Aspinwall at once landed, and remonstrated with the Swedish governor, who "acknowledged he had acted ill, and promised all favor." Both the Swedish and Dutch governors allowed them to proceed up the Delaware, but neither would allow them to trade, and each appointed a pinnace to attend theirs. "But the master of the vessel proved such a drunken sot, and so complied with the Dutch and Swedes, that they feared that when they had left the vessel to have gone up to the lake in a small boat, he in his drunkenness would have betrayed their goods to the Dutch, whereupon they gave over and returned home." The Swedish lieutenant made them pay forty shillings for the shot he fired at them before they left the river.[1]

Both the Swedes and Dutch were extremely jealous of the pertinaceous attempts of the English to settle on the Delaware; and both used their utmost

[1] Winthrop, vol. 2, p. 161, 179, 187.

endeavors to prevent them. They were between the English of Virginia on the one side, and New England on the other, both swarming with population. They knew that the English claimed the Delaware, and that if they once got a foothold they could not be expelled. Hence the not unnatural desire to prevent them making any permanent lodgment.

The ship Fame (or Renown), that brought Printz over, and the Key of Calmar, the first vessel that brought the Swedes to settle at Fort Christina, sailed from the Delaware for Sweden with valuable cargoes. From some cause they were compelled to put into the province of Friedland, where duties were demanded from them by the Dutch, as sovereigns of the territory in which New Sweden was situated. A long correspondence took place between the Swedish minister at the Hague, and the State General in regard to the claims of both parties to South River. The cargoes were finally liberated by the payment of an impost duty of eight per cent. under protest. These vessels had on board 2,127 packages of beaver, and 70,421 pounds of tobacco.[1]

On the 14th of October, William Penn, who received a grant of the State of Delaware from the Duke of York, and who was the founder of Pennsylvania, was born in London. Queen Christina also assumed the government of Sweden this year.

The English at Boston, undeterred by the bad success of the expedition under Mr. Aspinwall, again fitted out an expedition to the Delaware, which ended

[1] O'Cal., vol. 1, p. 371.

still more disastrously. They had procured a good supply of beaver, when some Indians who were trading with them suddenly drew out hatchets from under their coats, and killed the master and three others, and took away a man and a boy. The man saved was named Redman, and he understood their language. The Indians gave him a large portion of the goods. He lived amongst them for five or six weeks, when Printz got Indians to go and fetch him, and then sent him to Boston, where he was tried for betraying his companions, but acquitted.[1]

[1645] Jan Jansen Van Ilpendan, the Dutch governor on the Delaware, was, on the 12th of October, removed upon the charge of fraud, and Andreas Hudde appointed in his place. Jansen appears to have neglected the Dutch interests, and played into the hands of the Swedes, as instanced by his allowing his carpenter to assist in building for the Swedes the fort that cut off the Dutch trade.[2]

In December, Fort Gottenberg was destroyed by fire, and all the powder and goods blown up. It happened by a servant leaving a candle burning in the night, whilst he fell asleep.

[1646] This year commenced a series of disputes between the Dutch and Swedes, which finally ended in the overthrow of the Swedish power on the Delaware. Andreas Hudde was now the Dutch commissary or governor on the Delaware, a more active, energetic, and pertinaceous man than Jan Jansen, as well as far more faithful to the interest of

[1] Winthrop, vol. 2, p. 237.

[2] Albany Records. Acrelius.

his employers. Determined to get some of the trade of the Minquas on the Schuylkill, he ordered Captain Blancke, the commander of a sloop that had just arrived from Manhattan, to ascend the Schuylkill for the purpose of trading with the natives. Blancke was immediately warned off by the Swedes. Refusing to go, the next day Printz sent a minister of the gospel to Hudde (probably Campanius) who informed him that if the bark was in the Schuylkill, "he should compel her to leave it." Hudde claimed the right to trade in any part of the Delaware, and protested against any losses or hindrances that might arise from the proceedings of the Swedes. After an angry altercation between Hudde and some Swedish officers sent by Printz, in which Hudde remarked that "he would remain and see who would be so daring as to drive" him "away," Printz sent a letter to Captain Blancke, ordering him to leave the Schuylkill, "and seek the spot where sloops are usually accustomed to trade," under penalty of forfeiting both his vessel and cargo. The Dutch captain therefore left,[1] as he was ignorant of the causes of the dispute between the Dutch and the Swedes, and, being a private person, if his vessel and cargo were forfeited it would be a difficult task to recover them.

Hudde now infused great activity into the Dutch affairs, and appears to have made every endeavor to extend the Dutch influence and dominion.. He endeavored to ascend the Delaware above the falls, in obedience to instructions from Manhattan to look for

[1] See Hudde's Report, Albany Records, vol. 1, pp. 429, 451.

minerals, but was prevented by the Indians, who told him that the Swedes had informed them that the Dutch were coming from Manhattan with two hundred and fifty men to kill all the savages below the river, and erect a fort that would prevent the savages above from coming to their assistance. In proof of this they said the Swedes had also told them that the Dutch "would first come up in a small vessel and explore the spot, and then kill two savages as a pretext, but that Printz would never permit it." Hudde was therefore compelled to return.[1]

The Dutch governor at Manhattan, William Kieft, granted one hundred morgans of land opposite Reedy Island (called by the Dutch the little island of F. Vogelssant) in the neighborhood of what is now Port Penn, in New Castle county, to Abraham Planck, Simon Root, Jan Andriessen, and Peter Harmensen.[2] But the men never came and took possession.[3]

President Eaton, of New Haven, under date August 12, wrote a letter to Governor Kieft, of Manhattan, complaining of the outrages suffered by the English on the Delaware.[4]

On the 7th of September, Hudde received orders from Manhattan "to purchase some land from the savages situated on the *west shore, about a mile distant from Fort Nassau to the north*." In obedience to this order, on the 25th of the same month, Hudde purchased it from the Indian proprietor, and erected the arms of the Dutch West India Company upon it, upon

[1] Hudde's Report.
[2] Albany Records, vol. 1, p. 153.
[3] Acrelius, p. 417.
[4] Hazard's Historical Col., vol. 2.

a pole. As one Dutch is four English miles, this must have been part of the land upon which the city of Philadelphia now stands. The Swedish commissary, Hendrick Huygen, at once pulled down the the arms, which caused fresh protests to be made by Hudde. Printz also protested against the purchase, and claimed that the land belonged to the Swedes.[1] This brought forth a counter protest by Hudde, who complained grievously of the insulting manner in which the arms of the company were torn down, besides the many "bloody menaces" which were reported to him from time to time. This last protest was made on the 22d of October. It was sent to Printz in charge of a sergeant. Printz (the sergeant said) received it contemptuously, throwing it on the ground to one of his attendants, and saying, "take care of it." The attendant picked it up, and the sergeant was kept there waiting. Printz then departed to meet some Englishmen just arrived from New England. After some interval the sergeant asked to see the governor to obtain an answer, when he was thrown out of doors, and Printz (perhaps exasperated by the interruption of the sergeant) "took a gun from the wall to shoot him, as he imagined, but was prevented from his leaving the room."

This treatment Hudde complained of, as being very common on the part of Printz. "Freemen," he said, "as well as servants, when arriving at where he resides, are in a most unreasonable manner abused, so

[1] The Swedes had before purchased all the lands from the falls of Trenton to the Whorekill. See ante p. 153.

that they are often, on returning home, bloody and bruised."[1]

[1647] George Lamberton, who endeavored to settle on the Delaware, and with such bad success, was lost at sea whilst on a voyage to England.

Peter Stuyvessant arrived at New Amsterdam on the 11th of May as governor of New Netherlands in place of Kieft. He commenced his administration on the 17th of May. Shortly after his arrival he wrote complimentary letters to the Governor of Massachusetts and New Haven, desiring to be on friendly terms with them, but at the same time asserting the right of the Dutch to all the land between the Delaware and Connecticut rivers. The Dutch called the Delaware the South River, the Hudson the North River, and the Connecticut the Fresh River.

The quarrels between the Swedes and the Dutch on the Delaware still continued. The Armewsick savages, one day at noon, attempted to surprise the Dutch, but they by some means got information of the attack. This, and a misunderstanding amongst the savages, rendered the attempt abortive. Hudde insinuated it was by the machinations of the Swedes that this attack was made. "Printz," he says, "leaves nothing untried to render the Dutch suspected by both savages and Christians." Printz, both from English and Dutch accounts, appears to have been a violent abusive man. Upon Hudde urging the older claims of the Dutch to the Delaware he told him "*that the devil was the oldest possessor of*

[1] Hudde's Report, N. Y. Hist. Coll., vol. 1, p. 435, 436.

hell, but that he sometimes admitted a younger one." "This," said Hudde, "he declared at his own table, on the 3d of June, in presence of me and my wife, and many other equally vulgar expressions, serving and attended for the same purpose." Printz was also charged with tampering with the Maquas Indians, who lived near the Dutch possessions at Manhattan, (New York) and furnishing them with guns and powder.[1] Printz also stopped a Dutch vessel called the Siren, examined the goods, and tumbled them, and took from her a quantity of powder and shot.[2]

Several vessels arrived from Sweden this year with merchandise and settlers. They were the Swan, the Black Cat, the Key,[3] and the Lamb.

[1648] On the 20th of January, an ordinance was passed in Sweden, granting one-third of the excise on confiscated tobacco to the support of New Sweden, it having been found that the grant of the excise in 1642 did not produce half the sum expected. If this third of the excise did not prove sufficient, the balance was to be made up from the revenues of the crown. Goods from Holland landed at Gottenberg, but not intended to be sold in Sweden, were allowed to pass to New Sweden without duty.

The dispute between the Dutch and Swedes still continued on the Delaware, the Swedes being extremely arrogant, and appearing to pay no respect whatever to the Dutch or their authority. The

[1] Albany Records, vol. 3, p. 248. [2] Holl. Doc., vol. 8, p. 48.

[3] Probably the Key of Kalmer, the first vessel that brought over the Swedish settlers.

Swedes apparently desired to pick a quarrel with them, and drive them from the river.[1] A vessel, on the 2d of April, sailed past Fort Nassau without showing her colors, so that Hudde was doubtful as to what she was. Two guns were fired from the fort, but the vessel paid no respect to them. Hudde sent a boat with eight men after her, but the weather being fair, he was unable to overtake her. In two or three days she returned, with her colors flying. It proved to be a Swedish bark. Hudde then asked the skipper (Claret Huygen) why he "passed the fort without showing his colors, by which it might be known who the master was, though he had colors with him, as was evident, since they were now flying." The Swedish captain answered him very contemptuously, "that if he had known this would have come into consideration, he would not have done it now, but that he certainly should do so in the future, if it was to irritate, and a mark of his contempt." Hudde, who was extremely sensitive as to any thing concerning Dutch honor and authority, immediately sent a protest to Printz in relation to this conduct. In it he stated that it was contrary to an arrangement made between them, viz., that the vessels of each were to stop at the forts of the other, even when it was sufficiently known from where they came," Hudde very justly says, "so that neither your subjects and ours might be exposed to any mishap, whilst it was to be feared that otherwise, that under

[1] MSS. American Philosophical Society, Register of Penn., vol. 4, p. 315.

this cover, one or other foreign nation, to our great injury, might pass by."[1]

During the whole winter the Swedes in the neighborhood of Schuylkill had been gathering logs with the evident intent of building. Hudde, who kept a sharp watch over all their proceedings, supposed they were going to build near the place where the vessels usually laid at anchor. The correctness of his suppositions were soon made evident. On the 4th of April, some of the sachems from "Passayunk" called on Hudde, and inquired "why he did not build on the Schuylkill, as the Swedes had already done so." Upon inquiry, he found this to be the truth, and "in some places, too, of the highest importance." Accordingly Hudde commenced preparations to build too. He obtained a grant of land near the Schuylkill, in the neighborhood of Fort Nassau, and on the 27th of April went there with the necessary timber. He also called on the sachems who had granted him the land, and stated his intention of building on it. They sent a message "to the Swedes, who lived there already, and commanded them to depart, insinuating they had taken possession clandestinely, and against the rule of the sachems." That they (the sachems) had ceded it for the present to the Dutch," and that Hudde "should build there too." Then "Martt Hoock and Wissemenets, two of the chief sachems, planted with their own hands the colors of the Prince of Orange, and ordered that Hudde should fire a gun three times in token of possession." "This was done,

[1] Hudde's Report, p. 427.

and the house raised in presence of the chiefs. Towards evening," Henry Huygen, the "Swedish commissary arrived, with seven or eight men, and asked Hudde with whose permission or order he had raised that building." He answered, "By the orders of my masters, and the previous consent of the savages." Huygen asked if Hudde "could show any order of his master, and not letters of some freemen." Hudde answered, "Yes, and was ready to produce them, when he (Huygen) had shown by whose order he made that demand." "The sachems then told Henry Huygen that they should grant" the Dutch "that tract," and they "should settle there," and asked "by what authority the Swedes had built there," and "if it was not enough that they were in possession of Mattenehonk, the Schuylkill, Kingsessing, Kankanken, Upland, &c., possessed by the Swedes, all which they had stolen from them. That Minuit, now about eleven years past, had no more than six small tracts of land up Pagahacking, purchased to plant there some tobacco, of which the natives, in gratitude, should enjoy one half of the produce." They wished to know if the Swedes, "when purchasing one tract of land, should take those next adjoining it, as they had done, and yet continued to do."

They were greatly surprised that the Swedes, who had only lately arrived on the river, should prescribe laws to those who were the original and natural proprietors of the land, as if they might not do with their own as they wished. That the Swedes who had only lately arrived on the river had taken a great deal of land from them, whilst the Dutch had never taken any from them, though they had dwelt there for "thirty years."[1]

Such was the speech of the Indian sachems to the Swedes, at least so Hudde wrote to Stuyvessant at New Amsterdam, for we have no Swedish accounts. But it had no effect on their determination to stop the Dutch from building on the Schuylkill. Hudde commenced the erection of palisades around the house, because, said he, the Swedes had before destroyed the house which the company possessed on the Schuylkill, and built a fort in its place, and they perhaps might do the same here.

Whilst Hudde was thus engaged, "Moens King, lieutenant of the fort on Schuylkill, arrived," with twenty-four men, fully armed, with charged muskets, and bearing maces, and marching in ranks. The Swedish lieutenant "commanded his soldiers to lay down their muskets, and each take his axe in hand and cut down every tree which stood near and around the house." They "destroyed even the fruit trees" which Hudde "had planted there."[2]

This outrage of the Swedes was soon reported at New Amsterdam. The council on there, on hearing of it, sent S. Van Dincklage and La Montague to inquire into the affair. They arrived on the 7th of June.[3] They met the sachems on the 10th, and received a formal transfer of the land from the Indians.

[1] Hudde, p. 440. [2] Ibid.
[3] Ibid.

This land had fifteen years previously been transferred to the Dutch by the Indians in 1633. The conveyance was to Arent Corssen, the Dutch Commissary, that year at Fort Nassau, and was then called Armenveruis. The Dutch had erected Fort Beaversreede upon it.[1] The Indians had been paid in part, but not in total. This grant must therefore have been the final execution of the bargain previously made. After finishing this business they sailed from New Gottenberg, where they were received by Commissary Huygen and Lieutenant Papeogya, "who left them about half an hour in the open air, and a constant rain." After they were admitted to an audience with Printz, "they delivered him a solemn protest against his illegal occupation of the Schuylkill, to which he promised a reply before their departure."[2]

It is unnecessary to mention in detail the numerous disputes that took place between the Dutch and Swedes as they principally occurred in the country in the neighborhood of the Schuylkill. They belong more to the history of Pennsylvania than to Delaware. It will be sufficient to state, the Dutch continued to grant land on the disputed territory, that they several times attempted to build, but that they were in every instance driven off by the Swedes. And that Hudde, having no other mode to oppose Printz's violence, still pertinaciously protested, copies of which he always forwarded to New Amsterdam, and thus made the injuries matter of record, which record is

[1] Holl. Documents, p. 32, 50, 55, 56.

[2] Hudde's Report, p. 440.

now of the greatest service in elucidating the past history of this State.

The following are a few of the many outrages perpetrated: Hans Jackson made an attempt to settle in July. His buildings were destroyed by the son of Governor Printz, who threatened that if he came there again and attempted to build, he would give give him a "good drubbing." Thomas Broen attempted to settle at a place which he gave the name of New Holme. The Swedes, under Sergeant Gregory Van Dyck, pulled down his buildings, and told him if he did not leave they would beat him.[1] In this state of affairs Hudde left for Manhattan, when Printz, on the 16th of September, built a house in front of the Dutch fort of Beversreede, of twenty feet broad, and from thirty to thirty-five feet long, the back gable of the house being within twelve feet of the fort, entirely cutting it off from a view of the water. Simon Root and others endeavoring to build near Fort Beversreede were driven away by Lieutenant Swen Schute, and their buildings pulled down. Thus the quarrel went on, until the Dutch gathered up their strength, exasperated at the repeated injuries, and finally destroyed the Swedish power on the Delaware. Of which, however, more hereafter.

Two Swedes, who, with seven or eight guns, some powder and balls, went to trade with the Manquas, a tribe of Indians residing near Fort Nassau, were killed by them. This is the first recorded instance of any Swedes being killed by the Indians.[2]

[1] Hudde's Report, p. 440. [2] Albany Records, vol. 5, p. 71.

CHAPTER XVI.

FROM 1649 TO 1653.

English complaints laid before the United Colonies of New England—Their action—They write to Stuyvessant—Stuyvessant meets them at Hartford—An arbitration agreed upon—The New Englanders claim between the fortieth and forty-eighth degrees of north latitude—All the land claimed by the Dutch wanted by the English—Stuyvessant not ready with his proofs—They are in Holland—The arbitrators award both parties to settle on the Delaware on the lands they had purchased—Stuyvessant visits the Delaware—Endeavors to settle boundaries with the Swedes—Swedish claims—Their proofs in Sweden—Dutch allege Indians did not sell lands to Swedes—Stuyvessant meets the Indians—Buys from them the land between Fort Christina and Bombay Hook—Builds Fort Cassimer—The name—Supposed origin—Abandonment of Fort Nassau—Stuyvessant returns to New Amsterdam—Printz protests against the erection of Fort Cassimer—English from New Haven call at New Amsterdam on their way to Delaware—Stuyvessant arrests them—Threatens to forfeit their goods and send them to Holland—The English complain to the United Colonies—United Colonies promise them aid, if they send expedition—They accuse Stuyvessant of breaking the agreement—Write to their agent in London—Captain Mason applied to, to go to the Delaware—He declines—Commissioners from United Colonies visit New Amsterdam — Complaint of English against Stuyvessant—Stuyvessant threatens to prevent English settlement on Delaware by force of arms.

[1649] THE injuries done to the English on the Delaware by the Dutch, and their claims to the territory on that river, were brought to the notice of the United Colonies of New England at the meeting of the Commissioners at Boston, by Governor Eaton of New Haven. The Commissioners, however, declined to encourage, "by any public act," the settlement of the Delaware, as they could not spare the men, the English plantations in New England not having a sufficient supply of hands.[1] They, however, caused a letter to be written to Stuyvessant on the 15th of August, in which they (alluding to a former letter from the Manhattan authorities) stated that the answer they received in relation to the letter written, complaining of injuries received by the English on the Delaware Bay, was not satisfactory. They asserted the right of the English to the tracts on the Delaware, and that whilst the people of New Haven would neither encroach nor in any way disturb the peace of the Dutch, they must not fail in maintaining the rights and interests of the English there.[2]

[1650] To settle the difficulties in relation to Delaware, between the Dutch and English of New Haven, Stuyvessant agreed to meet the Commissioners of the United Colonies at Hartford. This meeting was brought about at the solicitation of the commissioners. Stuyvessant arrived at Hartford on the 23d of September. The correspondence between them it was decided should be in writing, to avoid the inconvenience of speaking. After settling as many points as possible in this manner, there were others that could only be arranged by disinterested

[1] Hazard's Coll., vol. 2, p. 127.

[2] Trumbull's Conn., vol. 1, p. 184.

friends of each party, therefore an arbitration was agreed upon. Two commissioners were appointed by each party. Stuyvessant appointed Captain Thomas Willett and Ensign George Baxter. The United Colonies appointed Simon Bradstreet and Thomas Prence. Stuyvessant gave his arbitrators full power to settle "any differences between the two nations," to end and determine them, as they "might deem just and right, *with power to enter into such terms of accord for provisional limits and leagues of love and union betwixt the two* nations in those parts as to them should seem just and right." The New England arbitrators had similar instructions.

The New Englanders claimed by their patent all the territory between the forty and forth-eighth degree of north latitude. In other words, all the territory between a point a short distance north of Philadelphia, and another at the mouth of Chaleur Bay, near the river St. Lawrence, including not only what is within the present limit of the United States, but also the present British possessions of Nova Scotia and New Brunswick. This was according to the grant of the North Virginia Company in 1606 and 1620. Between the thirty-eighth and fortieth degree of north latitude, north of the Potomac, had already been granted to Baltimore. Thus every foot of land professed to be owned by the Dutch in America was claimed as being granted by the English government to English citizens.

The arbitrators met, and the colonies of Connecticut and New Haven laid these complaints before them.

As most of the injuries suffered were done by Governor Kieft, Stuyvessant was not prepared to make answer to them. They therefore made no judgment on them until Stuyvessant could lay the matter before the West India Company. Stuyvessant, however, still asserted this claim of the Dutch to the Delaware, and protested against any other claim. The English also claiming the right, and as Stuyvessant advanced no proofs, all the commissioners could do was to award that both parties were to remain in *statu quo prius*, (that is, in the same state as they were before,) and that they were "to plead and improve their just interests on the Delaware for planting and trading," only that "all proceedings were to be carried on in love and peace." This award, which settled nothing and amounted to nothing, was duly signed by the four arbitrators, in the presence of all the commissioners, and Stuyvessant promised to abide by their decision.[1]

Stuyvessant then visited the Delaware, and attempted to have a settlement betwen the Dutch and the Swedes, as regarded their limits and boundaries. But this could not be done as though Printz "*determined the Swedish limits wide and broad enough*, yet without any justification or proper proofs," giving as an excuse that all the papers relating to the purchase of the lands were not at hand, but deposited at the chancellery at Stockholm. [1651]

Stuyvessant asserted, in his report to the Dutch West India Company, that this reply of Printz's was

[1] Hazard's Historical Collection, vol. 2, p. 171, 172, and 218.

not true, as he tried to purchase from the Indian sachem Waspang Zewan the lands that the Swedes were then settled on. But that the sachem refused to sell the land because the Swedes "had for a long time, and against his inclination, and with a forcible hand, kept possession of a part of said lands, without ever having given him the least consideration for them." "This," said Stuyvessant, "the chief declared verbally and in writing to the director, in presence of several reliable persons." The same sachem, Stuyvessant also asserted, gave the Dutch "authority, in a proper manner, to inherit and possess forever," all the land between Racoon Creek and the Minquas, or Christina, on both sides of the Delaware.

Stuyvessant, for further security, summoned to meet him on the 19th of July all the Indian sachems who lived near the Delaware, and all the owners of land in the neighborhood. He then questioned the Indians in regard to the land they sold to the Swedes. The Indians denied "ever having sold any land to the Swedes," as they pretended, excepting the ground on which Fort Christina was situated, and some ground around it for a garden to plant tobacco in. They then granted to the Dutch all the land between Fort Christina and Bombay Hook, (called by them Neuwsings,) on condition they should repair the gun of the chief Pemmennatta, when out of order, and give them a little maize when they required it.[1]

Stuyvessant immediately prepared to erect a fort to secure his newly acquired purchase. Accordingly,

[1] O'Call., vol. 2, p. 166.

on the spot where New Castle now stands, a short distance north of the present town, he erected Fort Cassimer. We have no knowledge of the reason of the name. The directors of the East India Company were themselves surprised at it when they were informed of it, as it is more Swedish than Dutch. It was probably from John, Duke of Casimir, a Swedish noble, who was instrumental in aiding the first Swedish settlement on the Delaware. Stuyvessant, after having thus fortified his purchases, returned to New Amsterdam, first abandoning Fort Nassau, and removing the garrison to Fort Cassimer, and having several conversations with Printz, "wherein they mutually promised to cause no difficulties or hostilities to each other, but to keep neighborly friendship and correspondence together, and act as friends and allies."[1] Printz, however, protested against the erection of the fort.

Supposing that according to the terms of the agreement with Stuyvessant, that their right to settle on the lands they had purchased on the Delaware was conceded, Jasper Graine, William Tuttell, and other inhabitants of New Haven and Sotocket, to the number of fifty, hired a vessel, and sailed from there for that purpose. On the 14th of September, on their way they stopped at New Amsterdam, and informed Stuyvessant of their intention. He asked to see their commission. It was shown to him, and he then refused to give it up, placed the master of the vessel and four others in prison, and refused to let them out

[1] Holl. Documents, vol. 8, p. 32–50.

until "they pledged themselves under their hands" that they would not go to Delaware, and at the same time informed them that if any of them were afterwards found there, he should forfeit their goods, and send them prisoners to Holland. He also, on the 11th of April, wrote to the Governor of New Haven, affirming the Dutch right to the Delaware, and threatening to prevent any English settlement there, "with force of arms and martial opposition, even unto bloodshed."

The ill-used Englishmen made prompt complaint of the Dutch treatment at the next meeting of the commissioners of the United Colonies, who were extremely indignant at the act of Stuyvessant. They passed a resolution affirming that "they did not think it meet to enter into any present engagement against the Dutch, choosing rather to suffer injuries and affront (at least for a time) than in any respect to seem to be too quick," but that if they should at any time within twelve months, at their own charge, transport one hundred and fifty (or at least one hundred) able bodied men, with arms and ammunition and vessels "fit for such an enterprise," and the same was approved by the magistrates of New Haven, then if, while they "carried themselves peaceably," they met with any hostile opposition from the Dutch and Swedes, and they required any further aid or assistance, then the United Colonies should furnish them with a sufficient number of soldiers for their defence, they paying the expense, and their lands there and trade with the Indians, to be answerable until it was paid. They also wrote a severe letter to Stuyvesant, complaining of his breaking his agreement with them, telling him he showed at their meeting "no just title to the Delaware, and asserting that, by the agreement made between the arbitrators, the English settlers were to be allowed to settle on their lands. They also wrote to Mr. Edward Winslow, the agent in London, complaining of the conduct of the Dutch, of the dishonor placed upon the English nation by submitting to such outrages, and of their duty to preserve English title to so considerable a place as Delaware." They also claimed that the Dutch should be compelled to make satisfaction to the English they had injured in their persons and estates, and requested information from him as "*to what esteem* the old patents for that place (the Delaware) have with the Parliament or Council of State, where there hath been no improvement hitherto made by the patentees; whether the Parliament hath granted any late patents, or whether, in granting, they preserve not liberty and encouragement for such as have or shall plant upon their formerly duly purchased lands."[1]

The people of New Haven, determining to maintain the rights to the Delaware, applied to Captain Mason, a man of known courage and military skill, to remove with them to Delaware, and take the management of the company. He was inclined to go, but his services at home being deemed essential by the General Court at Connecticut, they unanimously requested he would abandon all thought of going to

[1] Hazard's Historical Collection.

Delaware, to which he yielded, and the design was abandoned.[1]

But little was done by the English of New Haven or the United Colonies for two years, when, at the request of Stuyvessant, three commissioners were appointed by the United Colonies to visit New Amsterdam, but without any result. From some cause the commissioners left suddenly, greatly to the surprise of Stuyvessant. On the 2d of May, the New England commissioners again wrote to Stuyvessant, in which they reiterated all their previous complaints, and say "that to this day they have received nothing but dilatory exceptions, offensive affronts, and unpleasant answers, as well in the South River Bay, called Delaware, as upon the Fresh River, called Connecticut." This letter brought an answer from Stuyvessant, in which he says he could not admit of any settlements on the Delaware River as being contrary to his express orders, and that he done nothing in warning and preventing the settlers from New Haven from continuing their voyage to Delaware, but what was proper, and thus forewarn and advertise them from "all damage or bloodshedding," which might result therefrom. This letter showed to the United Colonies that Stuyvessant had thrown off the mask, and was determined that no English settlement should be made, if he could prevent it, on the banks of the Delaware.[2]

[1] Trumbull, vol. 1, p. 298.

[2] Hazard's Historical Collection, pp. 256, 260, 268, 270.

CHAPTER XVII.

FROM 1653 TO 1654.

Swedes request Dutch to take them under their protection—Stuyvessant refuses—Reason—Writes to Holland in relation to it—Permission to take them if they apply again—Printz returns to Sweden—John Pappegoya appointed Governor—Affairs of New Sweden placed in hands of College of Commerce—Jno. Amundsen Besh—Rising appointed governor—His salary—His instructions—Grant of land to Rising—Privileges to settlers by the College of Commerce—Rising embarks for New Sweden in the ship Aren, with two or three hundred people—Arrive in the Delaware—Attack and capture Fort Cassimer—Bikker, the Dutch commander, censured—He takes the oath of allegiance to the Swedes—Rising rebuilds the fort—The engineer Lindstroem—Dutch think of giving up Fort Cassimer—Decide to hold it—Rising makes a treaty with the Indians—Naaman's speech—The English write to Rising, informing him of their claim to land in Delaware—Population on the Delaware—Its increase—Rising wants a wife—Death of Chancellor Oxensteirn—Abdication of Queen Christina—Charles Gustavus, King of Sweden—Peace between England and Holland—Dutch capture the Golden Shark—Offer to restore her if Fort Cassimer is given up—Rising refuses—Correspondence between Rising and the United Colonies on English claim—Indignation in Holland at capture of Fort Cassimer—Stuyvessant commanded to expel Swedes from Delaware—He makes preparations to do so—Expedition to Delaware discussed in meeting of United Colonies—Two magistrates, Eaton and Neuman, propose to lead it—New Sweden has the monopoly of exporting tobacco to Sweden—expenses and garrison of New Sweden for 1655.

FOR some time before the building of Fort Cassimer, there had been no arrival of succors [1653] from Sweden, and the Swedes becoming discouraged, made a request to Stuyvessant for him to take them under his care, they at the same time promising to become subject to the Dutch. Stuyvessant, to prevent trouble that might ensue, declined to receive them, being unwilling to undertake so important a step without consultation with the powers at home.[1] He accordingly, on the 6th of October, wrote to the directors, who gave him permission to exercise his judgment. In their reply they said that "population of the country, that bulwark of every state, ought to be promoted by all means, so that the settling of freemen ought not to be shackled, but rather encouraged, by all honest means; all such, therefore, who are willing to obey our laws and statutes, ought to be protected in their rights of citizenship to the utmost of our power."[2]

Governor Printz, who had long been desirous of returning home, it is supposed either in October or November of this year, returned to Sweden. Historians differ in regard to the correct time of his departure. Some place it in the year 1652. But one paper at least is extant dated "New Sweden, October 1, 1653."[3] It is more than probable this was his last official act. A letter from Sweden, that arrived at Delaware after he had left on his return home, said "that they should extremely regret his immediate departure, before" they could make arrangements "in regard to his successor, and for the government of the country; if, however, this should be imposing too much upon him, and their service would be equally

[1] Holland Documents, vol. 8, p. 32. [2] Ibid. p. 154.

[3] See Plymouth Records, vol. 2, part 1, p. 87.

well performed by those whom he might leave in the country, he was at liberty to return home."

Printz was more than probably mortified at the strength of the Dutch whom he had before so imposed upon when under the command of Hudde, but whom, under a Stuyvessant, a soldier equally as fierce and determined as himself, he was bound to treat with respect, as instanced in his inability to prevent the erection of Fort Cassimer. Printz was ungentlemanly, unjust, and unreasonable, both in his treatment of the Dutch and English. Rudman says, that becoming weary of delay, and apprehensive of danger from the near vicinity of the Dutch Fort Cassimer, being only five miles from Christina, he went back to Sweden. The same writer also informs us that he had become unpopular by a too rigid authority. Printz, after his return home, was made governor of Joukeoping.

Printz appointed John Pappegoya, his son-in-law, to take charge of affairs. He ruled Delaware on the interim between the departure of Printz, and was the fourth governor of Delaware. His term of office was about eighteen months.

The "press of business and other obstacles" preventing the government of Sweden from regulating the affairs of New Sweden "as the utility of the company and the interests of the government demanded," the management of the affairs on the Delaware was placed in the hands of the "General College of Commerce." They accordingly, in November, commissioned John Anundsend (or John Anundsen Besh, as he is sometimes called) a captain of the navy, who was about to embark on board a galliott, to take command of the seamen in New Sweden.[1] His duty was to "consist principally in performing with fidelity and zeal all the duties that appertained to a captain of the navy in endeavoring to procure every species of advantage to the benefit of her majesty and the Company of the South; and should he, by the grace of God, arrive in New Sweden, to superintend carefully the construction of vessels, in order that they may be faithfully and diligently built, &c." He was to conform in all things to the instructions of the governor, and to receive such annual compensation "as his capacity of captain" would "entitle him to." In future he was "to count with certainty upon the favor of her majesty, and to obtain from the country a more elevated employment." The government was required to receive him with the rank of a captain in the navy.

The College of Commerce, on the 12th of December, also nominated John Rising[2] (by some called John Claudii Rising) under the title of commissary, to take charge of affairs in New Sweden. The letter of his appointment stated "to aid for a certain period our present governor" of New Sweden. He was therefore only intended as a lieutenant-governor, but as Printz had sailed before he arrived, and he had the full charge of Swedish affairs in this country, we may

[1] MSS. American Philosophical Society, Reg. of Pennsylvania, vol. 4, p. 374. It is not certain from the language whether Anundsen was to command the seamen of the galliott or of New Sweden.

[2] This name, like all the names of this period, is spelt many different ways.

consider him the fifth governor of Delaware. He was allowed one thousand two hundred dollars in silver per annum in addition to what he may receive from the Company of the South, also one thousand rix dollars for his equipment for the voyage. His instructions in effect were to be as follows:

He was told to fortify and protect a harbor that had been established; to extend the Swedish possessions on both sides of the river, as far as possible, without, however, causing any "breach of friendship with the English and Dutch." He was to induce, if possible, the Dutch to abandon Fort Cassimer, by "arguments and serious remonstrances," but "without resorting to any hostilities. It is better," says his instructions, "that our subjects avoid resorting to hostilities, confining themselves solely to protestations, and suffer the Dutch to occupy the said fortresses, than *that it should fall into the hands of the English, who are more powerful, and of course the most dangerous in that country.* But it is proper that a fortress be constructed lower down the river, towards the mouth; employing, however, the mildest measures, because hostilities will in no degree tend to increase the strength of the Swedes in the country, more particularly as by a rupture with the Dutch, *the English may seize the opportunity to take possession of the aforesaid fortress,* and become, in consequence, very dangerous neighbors to our possessions in said country." Thus early was perceived by Swedes, as well as the Dutch, the danger that the English would finally conquer and occupy the whole of the territory.

Rising's instructions also recommended "employing every means to facilitate commerce, as the most likely course to benefit the country, such as not exacting more than two per cent., or even allowing to enter free, merchandise arriving or departing, if sold to or to be employed in trade with the savages. The same was to be exempt from duty in Sweden, provided it belonged to Swedish subjects, but foreigners were to be compelled to pay a duty, and were not to be permitted to ascend the Delaware with their vessels, but were only allowed to trade with the company. The duties and excise levied on foreigners or subjects, were "to be employed for the defence of the country, and profit of the Company."

"Purchasers of land from the Company or savages, and becoming subjects," were "assured of being admitted into the Company," and enjoying all privileges and franchises. No one was, however, to enter without consent of the government.

Before Printz left the country he was to render to Rising a full account of the situation of affairs. He was to place all that related to military affairs and the defence of the country in the hands of John Amundsen, establish a council formed of the best instructed and most noble officers in the country, of which Rising should be the director, in such a manner that neither Rising in his charge, nor Amundsen as governor of the militia in his, should decide or approve of any thing, without directly consulting with each other. Printz for the present was to give them written directions for their guide. Should he remain in the country longer, he was to accept Rising as commissary and assistant counsellor, and in the council those whom he should think most proper. They also granted to Rising as much land as could be cultivated by twenty or thirty peasants.

Various privileges were also at the same time offered by the College of Commerce, for the more extensive settlement of the country. Swedes sailing in their own vessels, had the liberty of trading in the interior of New Sweden, "as well with the savages as Christians, and the Company itself, without paying any greater tax than two per cent." They also had the privilege of importing "merchandise procured in trade on the river into any port belonging to Sweden free of duty, but foreign vessels were prohibited from trading on the river with savages or others, but with the Company alone." All Swedes were allowed "to establish on the lands of the Company as many colinies as they may be able at their own expense to keep on said lands, and employ them on plantations of tobacco, or in any useful manner during certain years of franchises, and under good conditions." Those who purchased lands from the savages or the Company were to have the same forever. These and various other similar regulations were made by the College of Commerce, for the guidance of affairs in New Sweden.

Résing embarked for New Sweden from Gottenberg in the ship Aren, Captain Swensko. [1654] The precise day of his sailing is not known, neither is that of his arrival. It is more than probable he sailed in December, and according to his own account he arrived in the Delaware, or South River, as it was then called, a few days before the 27th of May. He brought with him, besides a body of officers and troops, Peter Lindstrom, a military engineer, a clergyman named Peter ———, and a number of settlers, in all numbering some two or three hundred souls. They sailed up the South River until they came near Fort Cassimer, and were perceived by the Dutch, on (as they allege) the last day of May. Gerritt Bikker, the commandant of Fort Cassimer, immediately sent Adriaen Van Tienhooven, accompanied by some free people, to see who they were. The next day they returned, informing the commandant that it was a Swedish ship with a new governor, and that they desired possession of Fort Cassimer, which they said was lying on the Swedish government's land. About two hours afterwards, Captain Swensko and about twenty soldiers were sent from the Aren in a boat. They landed at Fort Cassimer, and were civilly received on the beach, near the gate of the fort by Bikker, who supposed they would inform him, if they had any intention to commit hostilities. But contrary to this, they hurried through into the fort, (the gate being open,) and some immediately went to different parts of the bastions. They then demanded, at the point of the sword, the surrender of the fort as well as the river.

This transaction was so hurried as hardly to give time enough for the Dutch to send two commissioners on board the Aren to demand of Rising his commission, and some little time for consultation; but before the commissioners had got on board, there were two guns fired over the fort, charged with ball, as a signal. The Dutch soldiers were then deprived of their side arms, and aim taken on them ready to fire because they did not deliver up their muskets, which were immediately snatched from their hands. In the meantime the commissioners who had been sent on board the Aren returned, and brought information that "there was no desire to give one hour's delay;" that the commission of Rising was on board the vessel, and that the Dutch would immediately perceive the consequences of it. The Dutch soldiers were then chased out of the fort, their goods taken possession of, as was likewise the property of Bikker, the commandant. The Swedes confiscated every thing in the fort. Bikker, in his letter to Stuyvessant, giving an account of the affair, says: "I could hardly, by entreaties, bring it so far as to bear that I with my wife and children were not likewise shut out almost naked. All the articles which were in the fort were confiscated by them, even the corn, having hardly left as much as to live on, using it sparingly, &c. The governor pretends that her majesty has license from the state general of the Netherlands, that she may possess this river provisionally."[1]

This was the first fortress captured by civilized men on the Delaware. This treacherous capture of the Dutch fort, as will be seen in the sequel, was retalliated on the Swedes, by the capture of all their

[1] Holl. Doc., vol. 8, p. 85, 87.

possessions on the Delaware, and the overthrow of their power there forever.

Fort Cassimer was captured, according to Swedish accounts, on Trinity Sunday, and they accordingly named it Fort Trinity.[1] According to Dutch accounts, it was captured on the 31st of May.[2] The Dutch garrison only numbered some ten or twelve soldiers. Bikker was severely censured by Stuyvessant and the West India Company for his easy surrender of the fort. Stuyvessant spoke of it as a "dishonorable surrender," and the Company as "a cowardly and treacherous surrender." Bikker's account of the affair sent to Stuyvessant was unsigned by him. He remained with the Swedes, and took the oath of allegiance.[3]

Rising immediately after the capture of the fort, wrote to Stuyvessant, giving an account of the affair. In his letter he alleged "that it was a matter of too great importance to settle between him and Stuyvessant," but that "the sovereigns on both sides would have to compromise the matter." He also had the fort rebuilt and made much stronger, under the direction of Peter Lindstroem, the engineer. Lindstroem also prepared for the Swedish government a large map, embracing both sides of the Delaware River to Trenton. The original was destroyed with the palace of Stockholm, in 1697, but a copy pre-

[1] Acrelius, p. 114; Campanius, p. 82.

[2] The difference in dates is more than probable the difference between the old and new style. Therefore, old style, the fort would be captured on the 21st of May; new style, the 31st of May.

[3] Holl. Documents, vol. 8, p. 89.

viously taken still remained among the royal archives in 1834.[1]

On the 2d of June, Stuyvessant had made a proposition to the council at New Amsterdam to abandon Fort Cassimer. It was, however, decided still to continue the garrison there. The reason for the proposition, it is supposed, was the threatened appearance of danger from the English. The news of the Swedish attack on the fort had not then reached New Amsterdam, although it was in the Swedish possession.

The Dutch residing near Fort Cassimer had already taken the oath of allegiance to become subjects to the Swedish crown. Rising accordingly prepared to make a treaty with the Indians. On the 17th of June, a meeting was held at Printz Hall, on Tinicum Island, of ten of the Indian sachems or chiefs, and there "a talk was made to them," in which it was offered on behalf of the Queen of Sweden, to renew the ancient league of friendship that subsisted between them and the Swedes, who had purchased of them the land which they occupied. The Indians complained that the Swedes had brought much evil upon them, for many of them had died since their coming into the country.[2] A number of presents were made and distributed amongst them, on which they went out and conferred for some time amongst themselves,

[1] A copy of this map is in the possession of Thompson Westcott, Esq., author of a valuable series of articles on the history of Philadelphia, now being published in the Philadelphia Sunday Despatch.

[2] This year the Indians held a council to consider whether they should destroy the Swedes. See ante pages 81, 82, 83, 84 and 85.

and then returned, and addressed the Swedes. Their principal spokesman was a chief named Naaman, whose dominions were on the creek of that name. This (Naaman's) creek is the most northerly of our streams, and flows into the Delaware, a little more than a mile from the Pennsylvania line. Naaman made a speech, in which he rebuked the rest for having spoken "evil of the Swedes," and done them an injury, and told them he "hoped they would do so no more, for the Swedes were very good people." "Look," said he, pointing to the presents, "see what they have brought to us, for which they desire our friendship." So saying, he stroked himself three times down his arm, which among the Indians is a token of friendship; and afterwards he thanked the Swedes on behalf of the people, for the presents they had received, and said that "friendship should be observed more strictly between them than it had been before;" that "the Swedes and the Indians had been in Governor Printz's time as one body and one heart, (stroking his breast as he spoke,) and that thenceforth they should be as one head;" in token of which he took hold of his head with both his hands and made a motion as if he were tying a strong knot. And then he made this comparison, that "as the calabash was round without any crack, so they should be a compact body without any fissure;" and that if "any one should attempt to do any harm" to the Indians, the Swedes should immediately inform them of it, and, on the other hand, the Indians would "give immediate notice to the Christians of any plot against them, even if it were in the middle of the night." On this they were answered that this would indeed be a true and lasting friendship, if every one would agree to it, on which they gave a general shout in token of consent. Immediately upon this the great guns were fired, which pleased them extremely, and they said, "*Poo, hoo, hoo; mockirick pickon;*" that is to say, "hear and believe the great guns are fired." Then they were treated with wine and brandy. Another of the Indians then stood up and spoke, and admonished all in general, that they should "keep the league and friendship which had been made with the Christians," and in "no manner to violate the same," nor do them "any injury, or their hogs or cattle," and that if any one "should be guilty of such violation, they should be severely punished, as an example to others." The Indians then advised some Swedes to settle at Passyunk, where there lived a great number of Indians, that they might be "watched and punished if they did any mischief." They also expressed a wish that the title to the lands which the Swedes had purchased should be confirmed, on which the copies of the agreement (for the originals were sent to Stockholm) were read to them word for word. When those who signed the deed heard their names, they appeared to rejoice; but when the names were read of those that were dead, they "hung their heads in sorrow."

"There was then set upon the floor in the great hall two large kettles, and many other vessels filled with *sappaun*, which is a kind of hasty pudding, made of maize or Indian corn. The sachems sat by themselves, the other Indians all fed heartily and were satisfied."[1]

The English still persisted in their claims to the Delaware, and having heard of the arrival of Rising, at a court held at New Haven, July 5th, it was ordered that a letter should be written to him, informing of the "propriety which some of the New Haven colony have to large tracts of land on both sides of the Delaware, and desiring a neighborly correspondence with the Swedes both in trading and planting."

The Dutch and Swedish population on the Delaware was at this time, according to a letter of Rising to Sweden, dated the 11th of July, *three hundred and sixty-eight* persons. When Rising arrived, there were only seventy persons. So that in a few months, by various arrivals, the population had increased more than five fold. The same letter desired the officer to whom it is written to "*look out for a wife*" *for him.* Rising's residence was in the fort at Christina.[2] In the same letter he recommended John Poppegoya as Schuten. John Amundsen, who was appointed to command the militia on the Delaware, as well as the seamen, it is more than probable never came to Delaware, as there is no mention of him in any proceedings taking place there.

In August, Axel Oxensteirn, the Chancellor of Sweden, who did so much for the settlement of Delaware, died. Christina, Queen of Sweden, (who may be reck-

[1] This account is copied from Campanius, pages 76, 77, 78.

[2] Translated from a French MSS.

oned also Queen of Delaware,) abdicated the throne in favor of her cousin, Charles Gustavus. On the 16th of July, peace was celebrated between England and Holland, who had been at war. The hostilities between those nations were confined to Europe. Notwithstanding the matters in dispute, there were no hostilities between them on this continent.

The unprovoked assault of the Swedes on Fort Cassimer, and the capture of the same, was now retalliated on them by the Dutch. On the 27th of September, a Swedish ship, called the Golden Shark, commanded by Hendrick Van Elswyck, bound to South River, by mistake or ignorance of the pilot, or from some other cause, put into the North River, and got behind Staten Island. On discovering his error, the captain dispatched a boat to Manhattan for a pilot to take him to South River. Stuyvessant at once arrested the master and seized the vessel, and brought it up to New Amsterdam. The crew of the Shark (which was described by the Dutch as an old and leaky fluyt of forty to forty-five tons burthen) were allowed to stay on board the vessel, whilst Elswyck was sent on to the South River, to invite Rising to visit Manhattan, and arrange the difference between them. Stuyvessant agreeing that if Fort Cassimer was restored to the Dutch, that they in return would restore to the Swedes the Golden Shark and its cargo. The rudder was, however, taken from the vessel, and two Dutchmen placed on board. Elswyck accordingly went to the South River, but Rising refused to visit Manhattan, preferring to hold on to Fort Cassimer, and let Stuyvessant have the vessel.

The Commissioners of the United Colonies met at Hartford, and on the 23d of September addressed a letter to Rising, again urging the claims of the New Haven Englishmen to land on the Delaware. Rising, it appeared, had written to them on the 1st of August, in which he spoke of "a treaty or conference before Mr. Endicott, wherein New Haven's right was silenced or suppressed," and asserted the right of the Swedes to the "land on both sides the Delaware Bay and River from the Capes." "This," said the letter of the commissioners, "is either your own mistake or at least the error of them that so inform you. We have perused and considered the several purchases our confederates of New Haven have there made, the considerations given, acknowledged by the Indian proprietors under their hands, and confirmed by many Christian witnesses, whereby their right appeareth so clear to us, that we cannot but assert their just title to said lands."

In the meantime, Stuyvessant, who, with the exception of the seizure of the Golden Shark, had quietly endured the outrage of the seizure of Fort Cassimer by the Swedes, had received information in regard to that affair from Holland. The directors of the West India Company were greatly exasperated, and in a letter to him, dated November 16th, issued orders to Stuyvessant "to exert every nerve to revenge the injury," and not only to recover the fort, and "restore affairs to their former situation," but to drive the Swedes from "every side of the river." Only that those "who desired to settle under the Dutch government should be allowed to." They also desired Stuyvessant, if possible, to get the Swedes to settle in other places within the Dutch district, as they "would be more gratified if the borders of the river were settled by Dutchmen." "No means," said they, "ought to be neglected in case of success, which God may grant for its accomplishment, either by encouraging Dutch settlers, by bounties, or other more powerful luring motives." They promised him "succors, both in vessels, materials, and soldiers," and ordered him to "press any vessels into his service that might be in New Netherlands." They informed him that he had nothing to fear from any other enemy, being at peace with the English, so that he could "take all the soldiers at New Amsterdam." He was also instructed to accept the services of all persons who might offer their services," as the citizens of New Amsterdam were fully strong enough to protect that city during his absence on the expedition to Delaware. They also instructed him to apprehend Gerrett Bikker, the late commander of Fort Cassimer, who, "from documents and private information, they are compelled to conclude, had acted very unfaithfully and treacherously." That he should be punished, "as an example to others who had shared more or less in that shameful transaction."[1]

Agreeable to these instructions, Stuyvessant went silently though actively to work to prepare a fleet and armament. For some months previously there had been protesting by the Swedes and counter pro-

[1] Albany Records, vol. 4, pages 157, 159.

testing by the Dutch in relation to the seizure of the Shark, in which the title of both to the South River were again and again gone over. But this was now stopped, and whilst the Swedes were lulled into security, and made no preparations for defence, the Dutch were quietly and energetically pushing forward their armaments to expel them from the Delaware by force of arms.

The English in New Haven still continued to discuss at their meetings at the court at New Haven, and of the Commissioners of the United Colonies, their right to Delaware, and the best means of obtaining the same. On the 27th of November, the court at New Haven met, and amongst other things in relation to it discussed the propriety of purchasing the right of the English proprietors for the lands they claimed there. They agreed to take three hundred pounds for it. Several declared their willingness to go if they had the right leaders. It was proposed to two of the magistrates, Samuel Eaton and Francis Newman, who agreed to take the matter into consideration. Another court was held at New Haven, December 11th. Eaton and Newman both made similar answers as regards taking the lead of the colony to Delaware. Eaton answered that "it was necessary there should be some leader to such a work, but for his part, this (New Haven) jurisdiction having an interest in him, which he must have respect to, but if it appears that God called him thereunto, he should be willing." Newman answered that "if a meet number for quality and quantity were ready to go, he was willing to accompany them in the spring."[1]

Charles Gustavus, King of Sweden, (also, therefore, King of Delaware,) granted by a decree to the Swedish West India Company, on the 23d of December, the right of importing tobacco into Sweden. The decree states that it hoped that "not only New Sweden will be able to support itself and prosper, but also that our nation will have greater opportunity and facility for accustoming itself to the navigation and commerce of America."[2]

The following were the estimated expenses for New Sweden for the year 1655, viz.: One commandant, 75 silver dollars per month; one captain, 36; one lieutenant, 24; one ensign, 18; one sergeant-major, 15; three gunners, 8, each of whom is to have charge of the magazine in his redoubt; one corporal, 9; one drummer, 7½; thirty-six soldiers at 4 dollars each; one provost, 9; one executioner, 6; 3 priests, 10; one superior commissary who shall also be book-keeper, 20; one fiscal, who shall also be adjutant to the commander, 12; one barber, (surgeon,) 15; one engineer, who is also secretary, 12; one sub-commissary placed at the river Hoernkill, (Lewistown Creek,) 12; total, 550½ dollars per month, 6,606 dollars of silver per annum, or 4,404 rix dollars.[3]

[1] New Haven Records, pp. 160, 161.
[2] Register of Penn., vol. 5, p. 15.
[3] Penn. Register, vol. 5, p. 15.

CHAPTER XVIII.

Petition of Jno. Cooper and Thos. Munsen to General Court of New Haven—Desire of magistrate to accompany expedition—To loan them guns and powder—To assist with money—Answer of New Haven Court—Delaware and New Haven to be under one jurisdiction—Governor to reside one year at Delaware and one at New Haven—Second meeting of New Haven Court—Swedes supposed to be too numerous to allow of English settlement—Party to go and treat with the Swedes—English attempt to settle Delaware appears to be abandoned—Dutch make preparations for the conquest of the Delaware—Send ship-of-war Balance from Holland—French privateer hired—Day of fast appointed—Sailing of the expedition—They reach the Delaware—It captures Fort Cassimer—Terms of surrender—Those who took the oath of allegiance to the Dutch allowed to stay—Twenty Swedes take it—Rising considers the surrender dishonorable—Swedish detachment captured—Rising sends Elswyck to remonstrate with Stuyvessant—He endeavors to persuade him not to attack Fort Christina—Fort Christina besieged—Arrest of Swen Hook as a Spy—Capture of Fort Christina—Terms of surrender—Outrages by Dutch—Stuyvessant offers Fort Christina to Rising, who refuses it—Rising plundered—His departure—Swedish preachers expelled—The name of the State New Netherlands—Original names of rivers and places—Cooper's Island—First manufactory in Delaware—Names of Swedish families—Dwelling places of Swedes.

[1655] THE General Court met at New Haven on the 30th of January, again to take into consideration the matters in relation to Delaware, when a petition was presented from John Cooper and Thomas Munsen in behalf of some fifty or sixty that were desirous of settling in Delaware Bay. They desired that Samuel Eaton and Francis Newman, two of the magistrates, should be allowed to visit Delaware, and examine into affairs there, to see what chances of success there would be for an English settlement there; and that if they did not desire to pay their expenses, they should be paid by New Haven. They also requested that two great guns and powder might be granted them, that the lands of those going might be freed for a time from rates and public charges, and that a sum of money might be raised in that jurisdiction, "either to buy a small vessel that should attend their service, or otherwise, as should be thought meet." The court, in answer to the petition, allowed Newman and Eaton, and such others as might choose to follow them, to go to Delaware. Instructed them "either to take the propriety of all the purchased lands into their own hands, or grant it to such as should undertake the planting of it." Desired "that it should remain part of New Haven jurisdiction." Provided for its future government, by declaring that it (the English plantations on Delaware) should become greater in population than New Haven. That their "due consideration should be taken for the ease and convenience of both parts," so that the governor might be one year on the Delaware and another at New Haven. That the court for making laws should be ordinarily but once a year, and at the place of the governor's residence. That if the plantations should increase in Delaware and diminish in New Haven, that possibly in that case the governor might reside constantly in Delaware, and the deputy governor reside at New Haven. But the lesser part of the jurisdiction was to be promoted and eased by the greater part, "both in rates and otherwise." In relation to the lending or granting of any thing, they promised to propound it to the several plantations, "and promote the business for procuring something in that way."[1]

In the meantime John Cooper, one of the petitioners, went to Delaware, and a special court of New Haven was called on the 16th of March to hear his report. He stated that he found "little encouragement in the bay; but few were willing to engage in the settlement at present." During the debate on the subject, a Mr. Goodyear said, "Notwithstanding the discouragement from the bay, if a considerable number appear that will go, he would adventure his person and estate to go with them, in that design, but a report that three ships had come to the Swedes, seems to make the matter more difficult." After debating the case, it was voted that New Haven should be at twenty or thirty pounds charge, and that Mr. Goodyear, Sergeant Jeffery, and such others as they should think fit to take with them, might go to Delaware with a letter for the commonwealth, and "treat with the Swedes about a favorable settlement of the English on their own right; and then, after harvest, if things be cleared, the company may resort thither for the planting of it."[2]

Another court was held at New Haven on the 9th of April, in relation to this matter. Several citizens

[1] New Haven Records, pp. 83 and 85.

[2] New Haven Records, p. 165.

of New Haven being willing to go, it was ordered that they should still belong to the jurisdiction of New Haven. That their families should not be compelled to employ watchmen in their absence. That such of their lands and houses as lay unimproved, should be free from all rates for one year. They also agreed to lend them two guns belonging to the town, (if they could get permission of the jurisdiction for it,) half a hundred shot, a proportion of musket bullets, and a barrel of powder.[1]

This is the last account we have of the attempt of the English of New Haven to settle on the Delaware. Probably the great strength and energy exhibited by the Dutch in the overthrow of the Swedish power deterred them from carrying out their intentions. The United Colonies were not inclined or prepared to go to war with the Dutch for such purposes, and New Haven itself would have been too weak alone to engage in such an enterprise. The desire of New Haven, moreover, appeared to have been rather to have such a settlement made by the private enterprise of her citizens, than to engage in it as a community.

In the meantime, extensive preparations were being made both in Holland and Manhattan to recover Fort Cassimer, and overturn the Swedish power on the Delaware. In Holland, the drum was beaten daily for recruits. A ship of war, called the Balance, under command of Captain Frederick de Koninck, was sent over, as well as two or three vessels, amongst which

[1] New Haven Records, pp. 166, 167.

were the ships Great Christopher and King Solomon. Gunners, carpenters, and powder were also sent from Holland. In New Amsterdam, every exertion was made to procure volunteers. Stuyvessant being sick, Vice-Director Mathias De Stille, Attorney-General Tienhooven, and Captain Frederick de Koninck were appointed to superintend the fitting out of the expedition. A French privateer, named the L'Esperance, Captain Jean Flamand, that happened to be at New Amsterdam was also hired for the occasion. After hiring and pressing many vessels into the service that happened to be in the harbor of New Amsterdam, and getting supplies of men, provisions, and ammunition from others, the expedition, by the 5th of September, was prepared to sail. They had, however, previously, on the 25th of August, held "a general fast, thanks and prayer day," according to their language, to "implore the only bountiful God that it may please him to bless the projected expedition, only undertaken for the greater security and extension and consolidation of this province, and to render it prosperous and successful, to the glory of his name." The director and council prohibited "all usual exercises, as plowing, sowing, mowing, fishing and hunting on that day, and all other amusements and plays, all tippling and intoxication, under penalty of arbitrary correction."[1]

On Sunday, the 5th of September, accordingly the expedition sailed. It consisted of seven vessels, having on board six or seven hundred men. Stuyvessant was accompanied by De Stille, the vice-director,

[1] Albany Records, vol. 9, pp. 31, 32.

by Captain de Koninck, who commanded the naval part of the expedition, and the Rev. Mr. Megapolinsis, his chaplain. The next day (the 6th of September) they arrived in the bay of South River. The weather being fine, they met with no accident. Owing to "a profound calm and inconvenient tide, they did not reach Fort Elsingburg, which was then abandoned and in ruins, until the day following. Here Stuyvessant divided the force into five sections, each under its own colors. Making their preparations, took them some days. On the next Friday morning, they weighed their anchors and came opposite Fort Cassimer, about eight or nine o'clock, and anchored a short distance above the fortress. Stuyvessant sent Lieutenant Smith with "a drummer towards the fort, to claim the restitution of (to use the Dutch language) their *own property*."[1]

The Swedes had been informed some time before this, by the savages, of the intentions of the Dutch, and Rising had caused Fort Cassimer to be supplied with men and ammunition, to the best of his ability, and had instructed, in writing,[2] Captain Swen Schute, to defend the fort if it was attacked. He also ordered him to send on board these ships when they approached, and demand of them whether they came as friends, and in any case not to run by the fort, upon pain of being fired upon, (which in such case they

[1] We follow, in the account of this expedition, Dutch dates, which differs ten days from the Swedish.

[2] Stuyvessant's letter to the directors of the West India Company. Albany Records, vol. 13, p. 348.

16

could not reckon but as an act of hostility). On the contrary, if they were minded to treat with them "as friends concerning their territories and boundaries, he was to compliment them with a Swedish national salute, and assure them that they were well disposed to a fast friendship." Schute, however, disobeyed his instructions, and allowed the Dutch ships to pass the fort without firing a single shot.[1]

Upon the demand to surrender the fort, Schute requested time for consideration, and until he should communicate with Governor Rising. This was rejected. In the meantime, all the passes leading from the fortress to Christina were occupied by fifty of the Dutch soldiers. They even placed their forces as high up as Christina Creek. Schute was then a second time warned to surrender to save bloodshed. He answered this second summons by soliciting an interview for negotiation, which being acceded to, he met the Dutch in a valley about midway between the fort and a newly constructed battery. He then requested to be allowed to dispatch an open letter to Rising, after it was shown to the Dutch, which proposal was at once rejected. He then left the Dutch, dissatisfied, on which they approached the valley in sight of the fort.

In the meantime, after the Dutch had raised their breastworks to a man's height, the surrender of the fort was demanded for the third and last time. Schute (the Dutch say) then humbly supplicated a

[1] Rising's official report, New York Historical Coll., vol. 1, pp. 443, 448.

further delay, until early the next morning, which was granted, because they would not be ready with their battery, to approach yet nearer the fort under protection of their guns. The next morning, the 11th of September, Schute went on board the ship Weigh Scales or Balance, and signed the capitulation of Fort Cassimer on the following conditions, viz.:

"1. The commander shall, whenever it may please him, or when he obtains an opportunity, by the arrival of the *Croon*, or by any other private vessels, be permitted to transport from the Fortress Cassimer, all the artillery belonging to the crown, either large or small guns, as they were designed by the commissioner, 4 iron guns of 14 lb. balls, and five pieces, viz., 4 small and 1 large one.

"2. Twelve men, with their full arms and accoutrements, shall be permitted to march from the fort with the commissioner as his life-guard, the remainder only with their side arms, provided that the guns and muskets belonging to the crown shall remain at its disposal, or that of the commissioner, to transport them from the fortress whenever the commander may have an opportunity of bringing it to its execution.

"3. To the commander shall be delivered in safety all his personal property and furniture, which he may either carry with him or send for, when it shall please him, and so, too, all the goods of all the other officers, provided that the commander remains obliged to surrender, this day, the Fortress Cassimer to the director general, with all its guns, ammunition, and implements of war, and other effects belonging to the general privileged West India Company. Done, concluded, and signed by the combatants on the 11th of September, A. D. 1665, on board the man-of-war Weigh Scale (or Balance) at anchor in the neighborhood of Fort Cassimer.

"P. STUYVESSANT.
"SWEN SCHUTE."[1]

After the surrender of the fort, an offer was made to the effect that all persons who would take the oath were permitted to remain as freemen of South River, whilst those who had any conscientious scruples against taking it were to be allowed to dispose of their goods to the best advantage, and have a free passage out of the country. Twenty of the Swedes at once took the oath, of whom, however, only seven wrote their names. Swen Schute, the commander, also afterwards took the oath of allegiance and remained in the country. Those of the Swedish garrison that did not take this oath were transported to Manhattan.

Rising, in his official report of this affair, severely censures Swen Schute for the surrender of the fortress. He blames him for allowing the Dutch ships to pass the fort without firing a gun, whereby he says "they gained command of the fort and the whole river, and cut off communication between the forts by posting troops between them as high up as Christina creek. He also calls the surrender a "dishonorable capitulation," in which "he forgot to stipulate a place

[1] Albany Records, vol. 13, pp. 349, 350.

in which he, with his people and effects, might retire." He also complains of his "subscribing the capitulation not in the fort, or in any indifferent place, but on board a Dutch ship." But the world and impartial minds will justify Schute's surrender. The Dutch force was so overwhelming, that resistance was hopeless, and it could only have resulted in the shedding of human blood, without any possible benefit. Rising himself afterwards surrendered with the main forces to the Dutch with scarcely any more resistance than that made by Schute. The only hope of the Swedes to have defended themselves successfully, would have been in the concentration of their forces at some one point, either at Christina, Cassimer or Tinicum, where they might for a time have successfully withstoood a seige, and possibly tired the Dutch out, who were not prepared for operations that required length of time. But it is very doubtful if even this course would have been successful, so great was the power of the Dutch in comparison with that of the Swedes. The whole population of the Swedes the year before, including Hollanders, (who of course could not be relied upon to fight against their own countrymen,) men, women, and children, only numbered 368 persons, whilst the Dutch force numbered between 600 and 700 armed men.

Fort Cassimer had fallen so suddenly that Rising, ignorant of its surrender, had sent nine or ten of his best men to strengthen the garrison. This detachment crossed the Christina Creek early in the morning of (according to Swedish accounts) the 1st of

September, at the place, a few years since, known as the Old Ferry, but at present occupied by the Diamond State Iron Works. As soon as they arrived at the opposite bank, the Dutch, to the number of fifty or sixty strong, attacked them, and summoned them to surrender, but they put themselves in a posture of defence, and, after skirmishing with the Dutch, were all taken prisoners, except two, who retreated to the boat, and were several times fired upon by the enemy without being killed or wounded. Upon this the Swedes fired upon the Dutch from the fort, whereupon they retired into the woods, and afterwards harshly and cruelly treated such of the Swedes as fell into their hands.

The same day the factor Elsnyck was sent down by Rising from Fort Christina to Stuyvessant to demand an explanation of his conduct, and to dissuade him from further hostilities, as Rising asserted he "could not be persuaded that Stuyvessant seriously purposed to disturb the Swedes in their lawful dominions." Stuyvessant at first threatened to detain Elswyck as a spy. However, upon his asking him the reason and intention of the arrival of the Dutch, "with the orders of the principals," Stuyvessant informed him the Dutch "claimed the whole river and all the Swedish territory thereon." He then requested the Dutch to remain satisfied with Fort Cassimer, and endeavored to dissuade them from advancing any further on Swedish territory, or prosecuting any further hostilities against Fort Christina, using "extremely courteous language, now and then intermixed with threats." But Stuyvessant was neither to be frightened nor persuaded from his purposes, and accordingly Elswyck went back to Rising, and Stuyvessant prepared to move with his fleet and army to besiege Christina.

When Rising was informed of the ill success of Elswyck's mission, he collected all the people he could for the defence of Christina, and labored night and day in strengthening the ramparts and filling gabions.[1] The next day (the 2d of September[2]) the Dutch showed themselves in considerable strength on the opposite of Christina creek, but attempted no hostile operations. On the morning of the 3d they hoisted their flag on a Swedish shallop which lay drawn up on the beach, and established themselves in a neighboring house. Rising then sent Lieutenant Swen Hook with a drummer to demand "what they purposed," and for what purpose they posted themselves there, and (as if there could be any doubt on the the matter) as to whether the Swedes should view them as friends or enemies. When he had nearly crossed the creek, he asked the Dutch whether he might "freely go to them?" They answered, "Yes." And whether, after "discharging his commission, he might freely return." To which also the Dutch answered "Yes." So the drummer rowed the boat ashore, without beat of drum, (as a drummer could not well row a boat and beat a drum at the same

[1] Baskets filled with earth.

[2] We are now following Swedish records, and have to give Swedish dates.

time). The lieutenant was met by an officer, and conducted to a house at some distance off, which the Dutch had taken possession of. He was then sent down to Stuyvessant, who, believing him to be a spy, arrested him, and threw him into the ship's hold. Thus asserted Rising: "They treated our messengers contrary to the laws and customs of all civilized nations."

On the 4th of September, the Dutch planted gabions about the houses on the opposite shore of Christina Creek, in the neighborhood of where the Townsend Iron Works are now, and afterwards threw up a battery under cover of them, and intrenched themselves there. Rising supposed this indicated, as the purpose of the Dutch, to "claim and hold all the territory from Bombay Hook to the south bank of the Christina," which the Dutch had some time purchased of the Indians, and that they intended to construct a fort there to hold those possessions. Rising did not believe that the Dutch would commence hostilities against him until they had made some claim or promulgated some protest, as he had "received from them neither message nor letter assigning any manner or cause of complaint." Rising appears to have forgot that he himself had set the example of commencing hostilities, without warning, by his attack and capture of Fort Cassimer.

On the 5th they sent their transport ships up the Brandywine, then called the Fish Kill, or Fish Creek, to Tredie Hook, which means the third point or promontory. This was an elevated piece of fast land on the north side of the Brandywine, below the present railroad bridge, and below a brick house, which is still standing, and which used to be known as Vandever's brick house. Here they landed their men, and marching up the point a short distance, crossed the low valley which at that time was overflowed at flood tide, and made what was until a few years ago the farm on the eastern side of the railroad bridge, an island. It was called "Timber Island," and was formerly in the tenure of the Vandever family.[1] It was a few years ago purchased by the Christiana Improvement Company, who have divided it mostly up into lots, and laid out streets through it. Claymont and Heald, and Tenth, Eleventh, Twelfth and Thirteenth streets, when they cross the Brandywine, now cover its site. The army having now got possession of this island, passed over the west side of it, and threw up a battery, on which they planted four cannon, all pointing towards Fort Christina, and within effective gunshot of it. Leaving there two companies of troops to man the battery, they marched up the creek, and then crossed over the "Stoor Fallet" (Great Falls) meaning the Brandywine, more than probable at the place where the old ford used to be, at the foot of the old King's Road, below the first dam across the Brandywine, and which road is yet used, from Delaware Avenue to that stream, forming the western boundary of the cemetery.

[1] A number of the descendants of this family now reside amongst us. It is one of the oldest in Delaware. They are descended from one of the old Dutch settlers.

Having crossed the Brandywine, they marched eastwardly again, until they came to a place between the burying ground of the Old Swedes Church, and the Diamond State Iron Works, about what is now the neighborhood of Church and Fourth streets. At that time this was a high bank, and here they erected a battery, which they mounted with three cannon. They also erected another, within the present grave-yard of the Old Swedes Church, near its south wall. These batteries were manned with four companies. They also planted a battery of five cannon within the village of Christinaham, whose site was more than probable on what is now covered by the lower portions of Sixth, Seventh, and Eighth streets, and Buttonwood street and the railroad, where they come in juxtaposition. This battery was placed immediately behind the fort, in which they placed the largest body of their force, consisting of six companies. This was formerly high ground, but it is now leveled and almost entirely covered with buildings. They also had a battery erected on the fast land, on the opposite side of the Christina, of four guns, manned by three companies. This was about where the Townsend Iron Works now stand, at what used to be the old ferry, on the southern shore of the Christiana, where travellers crossed the Christiana before the erection of the present bridge at the foot of Market street.

Thus the fort was now invested on every side, except towards the southeast, in which direction there was nothing but low morass, which at high tide lay four or five feet under water. Through this low morass the channels of the Christina and Brandywine then pursued a serpentine course, but in different directions, but at length were united in one stream about half a mile southeast of Fort Christina. That the investment might be complete, the Dutch now brought up their armed ships, and anchored them in the mouth of the Brandywine.

The Dutch had about finished their preparations, without any molestation from the Swedes, when the latter burnt a little powder in a couple of their guns to scale them. The Dutch fired several shots over their heads from Timber Island, where they had taken post in a house. They also announced to the Swedes they had taken up a position on the west side by several volleys. Rising continued to make the best defence which his strength would allow, if he should be attacked, for he was not yet satisfied as to the intention of the Dutch, when a circumstance occurred that left him no longer in any doubt, for an Indian arrived with a letter from Stuyvessant, in which he claimed the "whole river," and required Rising and all the Swedes either to evacuate the country, or to remain there under Dutch protection, "threatening with the consequences in case of refusal." Rising answered by letter "that he would reply to this extraordinary demand by special messengers," and sent him back his answer by the same Indian. He then held a general council of war as to what should be done if the Dutch assaulted him by storm or battery, at which it was determined that the Swedes should in any case remain on the defensive, and make the best resistance they could, but not to commence or provoke hostilities, on account of their weakness and and want of supplies. That they should wait until they were fired upon, or the Dutch began to storm their works, and then defend themselves as long as they could, and leave the consequences to be redressed by their superiors.

The Dutch now began to make still further encroahments upon the Swedes. Rising, in his official report says: "They killed our cattle, goats, swine, and poultry, broke open houses, pillaged the people without the sconce of their property, and higher up the river they plundered many, and stripped them to the skin. At New Gottenberg they robbed Mr. Poppegoya's wife of all she had, with many others who had collected their property together there." The Dutch in the meantime continued to advance their approaches to Christina, which was a small, feeble work, and lay upon low ground, and could be commanded from the surrounding heights, all of which hostile acts, injuries and insults we were, to our great mortification, compelled to witness and suffer, says Rising, "being unable to resist them by reason of our want of men and of powder, whereof our supply scarcely sufficed for a single round for our guns." Rising, however, seeing he was unable to make any defence, still determined to try negotiation. He sent messengers to Stuyvessant, (who was staying at Fort Cassimer,) with a written commission, to dissuade him from further hostilities, again protesting against the invasion of Swedish territory, without assigning causes, or denying as far as they could the Swedish right to the river. Rising also suggested to Stuyvessant the jealousy that would undoubtedly ensue between their respective sovereigns, and other momentous consequences that might arise from his acts, as the Swedes were determined to defend their right to the utmost of their strength. That Stuyvessant would have to answer for all the consequences that might ensue. The message finally required him to "cease hostilities and retire with his people from Fort Christina." But all this availed nothing. Stuyvessant had received his orders from Holland to take the Swedish fortresses, and he was determined to obey them. So he simply answered Rising's letter by reaffirming the Dutch right to the Delaware; spoke of the Swedish claim as a usurpation, and never for an instant relaxed in his preparations to take the fortress.

The garrison of Rising consisted only of about thirty men. With this small force he was unable to make any sorties to prevent the Dutch from taking any position they desired. They had command of the fort so completely, that not a man could stand on the ramparts with security. Besides, some of the Swedes were sick, some ill-disposed, some had deserted, provisions were scanty, and all the men nearly worn out with watching. From these considerations, and from fear of a mutiny, when Stuyvessant, who had heretofore been at Fort Cassimer, but had now came up to Fort Christina, peremptorily summoned him, on the 12th of September, to capitulate, with the threat of giving no quarter, the Swedes proposed to Rising to go out and have another conference with Stuyvessant, and endeavor to "bring him to reason." On the 13th of September, accordingly, Rising, accompanied by the factor Elswyck, went out to the most advanced work of the Dutch. He was met by Stuyvessant and De Stille. The place of meeting was then an elevated spot behind the fort, but was cut down and levelled with the grade of the street, to fill up wharves constructed on the western side of where the fort once stood. Here, in presence and in full sight of both the hostile combatants, the representatives of the Swedes and Dutch commenced their parley.

Rising again protested to Stuyvessant against his proceedings, and again went over the argument in favor of the Swedish claim. But all this, as might be expected, produced no impression on Stuyvessant, who would listen to nothing but the surrender of the fort and the whole river. To this Rising replied that he would defend the fort to the last. The conference was then broken off, and Rising and Elswyck retired within the ramparts, and encouraged his men to make as manly a defence as they were able.

On the 14th, the Dutch having completed their works, "formally summoned Fort Christina with harsh messages, by drummer and messenger, to capitulate within twenty-four hours." Rising then assembled a general council of the whole garrison, who, as there was a want of powder and munitions, and no hopes of relief, decided unanimously to surrender, which was therefore done on the 15th of September, (old style, but new style, or present mode of reckoning, the 25th of September,) on the following conditions, viz:[1]—

1. All guns, ammunition, implements, victuals, and other effects belonging to the crown of Sweden and South Company which are in the fort and its vicinity, shall remain in full, property to the crown and company, whilst it shall depend upon the Governor either to take all these with him, or deliver them to the Director General, Peter Stuyvessant, on condition that all, when demanded, shall be returned without any delay whatever.

2. Governor Rising, with all field officers, and subaltern officers, ministers and soldiers, shall march out of the fort with beating of drums, fifes, and colors flying, firing matches, balls in their mouths, with their arms, first to Timmer Isle, (Building or Timber Island), where they all, at their arrival from the fort shall be lodged in the houses, with security, till the departure of the Director General with the man-of-war, the Weigh-Scales, which shall convey, at longest, within fourteen days, the Governor, with his people and goods, so far as the Sand Point, about five miles from Manhattan, in full security. Meanwhile it is permitted to Governor Rising and his factor Elswyck, with four or five servants, to remain so long in their houses in the fort till they may arrange their private affairs.

[1] The account of the surrender of Fort Christina is derived from Rising's official statements. See N. Y. Hist. Soc. Coll. vol. 1, pp. 443, 448.

3. All papers, letters, documents, and acts belonging to the crown of Sweden, the South Company or private individuals, discovered and obtained in Fort Christina, shall be delivered unopened and unsearched to the late governor, to be distributed at his discretion.

4. No person belonging to the crown of Sweden, or the South Company, officers, soldiers, ministers or freemen shall not be compelled to stay, but permitted to accompany the government wherever they may deem proper.

5. All the high and low ministers of Sweden or of South Company, ministers, officers, soldiers and freemen shall be maintained in the undisturbed possession of their individual property.

6. If any servant or freeman desires to leave this country, and now could not be ready to depart with the Governor and his people, all such shall be permitted to dispose of their real and personal property during the period of one year and six weeks, provided they take the oath of loyalty for the time they intend to remain on this river.

7. If there are any Swedes or Fins who do not wish to depart, then it shall remain free to Governor Rising to admonish them, and if they upon such admonition are inclined to follow him, then all such shall not be prevented by the Director General from doing so, while they who voluntarily are resolved to remain, and desire to search for their sustenance in this country, shall enjoy the liberty of the Augsburg Confession, with a minister to instruct them in this doctrine.

8. The Governor, John[1] Rising, the Factor Elswyck, with all the other high and low officers and soldiers and freemen, who now wish to depart with their personal property, shall obtain from the Director General a convenient vessel, which at Sand Point shall take them in, and transport them further to the Texel, and from there further with a *Boeÿer*, galliot, and other good ships, to Gottenberg, free from expense, provided that such ship or galliot shall not be molested or detained at Gottenberg, for which the aforesaid Governor remains responsible.

9. If Governor Rising, Factor Elswyck, or any minister of the aforesaid crown or South Company have contracted any debts, they shall not be subject to arrest within the government of the aforesaid general.

10. Governor Rising is permitted to inquire unmolested, how the late Commander Schute, officers and other soldiers have conducted themselves in the surrender of the fortress at Sand Point.

11. Provided the Governor engages to march out of the Fortress Christina on this day, being the 25th of the month of September, with all his men, and to surrender it to the Director General. Done and signed on 25th of September aforesaid, on the paved place between Fort Christina and encampment of the Director General.

PETER STUYVESSANT.

JOHN RISING,

Director of the Country of New Sweden.

[1] In some documents he is called Jno. Clandii Rising. In others. John Rising. His name is also spelt Risingh.

17

There was another article of this treaty, by which it was agreed that the skipper, with whom Governor Rising and the Factor Elswyck shall depart, shall be expressly commanded to land them either in England or France, and that Stuyvessant should advance to Rising £300 Flanders, which sum Rising agreed to pay in six months, at Amsterdam. In the meantime, the property surrendered to Stuyvessant was to be forfeited if the money was not paid at the stipulated time. The money was never paid, and therefore Stuyvessant kept the goods. Thus, after about seventeen years' rule, fell the Swedish power on the Delaware.[1]

After the surrender, the Dutch were accused of committing many outrages on the inhabitants. Some writers affect to disbelieve them, but the evidence is such as leaves but little doubt of their truth. Rising, in his remonstrance to Stuyvessant, accuses the Dutch of "plundering Tennakong, Uplandt, Findlandt, Printzdorp, and several other places." At Fort Christina, he says, "women were violently torn from their houses, whole buildings destroyed, and they dragged from them: yea, the oxen, cows and swine, and other creatures were butchered day by day; even the horses[2] were not spared, but wantonly shot, the plantations destroyed, and the whole country left so desolate that scarce any means are remaining for the

[1] The articles of surrender are from Dutch authorities. See Albany Records, vol. 13, pp. 353, 359. They of course have new style dates.

[2] This is the first mention of horses on the Delaware.

subsistence of the inhabitants." Acrelius says "the Swedes suffered great hardships from the Dutch. The flower of their troops were picked out and sent to New Amsterdam; though under the pretext of their own choice, the men were forcibly carried aboard the ships. The women were ill treated in their houses, their goods pillaged, and the cattle killed. Those who refused allegiance were watched as suspicious."

After the surrender, a tender was made of Fort Christina to Rising, but it was refused, the Dutch say, "under pretence that the affair was not complete, and he would rather hold himself to the capitulation made."

On the evening of the 28th of September, a mob of Dutch assaulted Rising and some of his men in the fort, and plundered them of all the goods they could find, much of which was Rising's private property. Rising and Elswyck were finally conveyed to Manhattan in the man-of-war Balance, when a sharp correspondence took place between Rising and Stuyvessant, Rising accusing the latter of breaking the terms of the capitulation in several particulars.

There is but little doubt that the Swedes after their surrender were disgracefully plundered, although it was probably against the wish or desire of Stuyvessant.

Rising and the rest of the Swedes finally took their departure for Europe in the ships Spotted Cow and Bear, but were compelled to put into England, where he gave the first information of the overthrow of the Swedish power on the Delaware, to the Swedish Minister. Thus terminated the short career of this rash and injudicious man, who in defiance of his instructions commenced a war which ended in the overthrow of himself and nation on this continent. The last heard of Rising was a report made to the Dutch authorities that "Hendrick Huygen had said that in September, 1660, Rising had been arrested in Sweden, but that after long entreaties it was at length consented to grant him an armed vessel for the recovery of South river."

"Two out of the three Swedish clergymen, it is said, were expelled the country; and the one left was of godless and scandalous life."[1]

The Swedish rule thus being ended by their being conquered by the Dutch, New Sweden ceased to be the name of our State. From the fall of the Swedes until its capture by the English, it was part of the Dutch territories of New Netherlands, and consequently went by that cognomen. The boundaries of New Netherlands were never accurately defined. The Dutch, however, at this time exercised jurisdiction over the settled parts of the State of Delaware, (which extended from Lewes to our northern circular boundary at the Pennsylvania line), and up the Delaware as high as Trenton, on both sides of that river, over the whole of the State of New Jersey, and both sides of the North, or Hudson river. They claimed the right of jurisdiction over the Connecticut, which they called the "Fresh river," but the English citizens of New Haven held it in conjunction with them,

[1] O'Call. vol. 2, p. 290, 318.

and finally expelled them. They also struggled with them for the possession of Long Island. New Netherlands, of which our State was then a part, may be said to consist of the State of Delaware, Pennsylvania, New Jersey, and New York, and part of Connecticut. They named the great rivers of this territory as follows:—The Connecticut, the Fresh river, (from being composed of fresh water); the Hudson river, the North river; and the Delaware, the South river.

The following were the names given by the Swedes to the various streams and places in our State. They are derived principally from the map made by the Engineer Lindstrom,[1] who has set them down both in Swedish and French. This map takes in the Delaware from Trenton to the Capes. The first name of a place is the Swedish, the second the French. Naaman's creek is named "Naaman," (the same as now), and "La Revier de Naaman." The creek to the South of Naaman's, and which we believe is the creek which flows into the Delaware near Hollyoak Station, on the railroad, was named "Naaman's Fallet," (or Falls), or the "Cataract de Naaman." The land between Naaman's and this creek, on the Delaware shore, was named "Windrufwe Udden," "Le Cap des Raisins." Udden appears to have been the Swedish name for "Cape." The French translated into English means Cape of Grapes. It is more than probable that grapes in the early settlement of the State grew

[1] We are indebted to Thompson Westcott, Esq., of Philadelphia for the perusal of this map.

there in great plenty, and probably "Windrufwe" is Swedish for grape. Shelpot creek was named "Skilpaddle Fallet," or "La Cataracte des Tortues," or in other words, Turtle Falls. This was first corrupted into "Skillpot," which was the name it bore in the time of Penn, and afterwards into Shelpot, its present name. The Brandywine was called the "Fiske Fallet," or "La Cataracte des Poisons," in other words Fish Falls, or Fish creek, or river. Cataract or Fallet being applied to its rougher parts. The Shelpot and the Brandywine then, according to the map of Lindstrom, had mouths, which carried their streams into the Delaware, as well as into the Christiana. The Brandywine, it is alleged, received its present name "in consequence of the loss of a vessel in its waters laden with brandy," in the Dutch language called Brand-wijn. This account is confirmed by the report of a number of old persons, who declared that the remains of the vessel were frequently pointed out to them in early life by their ancestors, with the assurance that those remains were parts of the ship whose loss gave rise to the name of the river.[1]

At the mouth of the Shelpot and Brandywine was formerly an island named "Rylflacht," or "Plaines des Rosseau." This was a beautiful little island, containing a few acres of land, studded with lofty forest

[1] See Ferris, p. 196. He also says: Frederick Craig, a very worthy citizen, and remarkable for a clear, retentive memory, who lived to eighty-five years of age, and died in 1841, has been frequently heard to say that the wreck of the ship which gave a name to the Brandywine, had often been pointed out to him in early life by the old people of that day.

trees, and called by the Indians "Manathan." This spot was inhabited in the time of the Swedes by two Dutch families, with their workmen, who followed the business of ship carpenters, boat builders, and coopers. Here they built yachts, a kind of fast sailing trading vessel, with which they could run up the creeks and inlets along the shores of the river, to trade with the natives. They also built boats and galleys, and made casks and tubs. From this it was named "Coopers' Island." This was the first workshop or mechanical establishment we have any record of as being carried on in this State. They occupied the island, according to Campanius, when Governor Printz first came over. After the country became more thickly settled, they abandoned their solitary home to reside amongst the planters. Being left again to the care of nature, unassisted by man, the cleared parts of the island were soon covered with the wild plum, or mountain cherry, which grew up spontaneously, and was the cause of its taking the name of "Cherry Island." By the embankment of the meadows, the island lost its insular character, but the name was retained, to give a distinctive appellation to the land around it, which is still called Cherry Island Marsh.[1]

At the mouth of the Christiana there was also a marshy island, on one side of which the Brandywine flowed into the Christiana, and on the other into the Delaware. This was also embanked, and has lost its name in that of Cherry Island Marsh. Amongst other names, it was then called "Rylflaett's Plaines,"

[1] See Ferris' Original Settlements.

or "Campagne des Rosceaux sont Marques pas des Points." The ground on which the main part of the city of Wilmington stands, was named "Hiort ad Cap de Corp." Clements creek, a small stream which flows into the Christiana west of the Harlan & Hollingsworth Works, near Justison street, was called "Liblefals Kylen," or "La Reveire de la Petit Cat-racte." There was a large island in the Christiana, in the neighborhood of where the town of Newport is now, which was called "Nootebohms," or "Ile de Codraie." The land between the Clements creek and the White Clay creek, on the north side of the Christiana, was called "Huis Kakamense," while Bread and Cheese Island, near Stanton, had the name of "Kees und Tyrodts a eller Rodoleaus oen ke da Fromage et du Paine," or in the main, the same names it has now in the English instead of in the French and Swedish language. The Christiana, before its junction with the White Clay creek, was called "Sickpeckons Sip-punck," or "Tasswagers Kyl," whilst the land on the south bank of the Christiana from the city of Wilmington to where the White Clay creek flows into it was named "Tura Udden," or "De Cap des Pins." On the same side of the Christiana, but east of Fort Christina, it was named "Kojaca Salung." It was also called "Tenacongs." On the river shore of New Castle hundred, between what is now the town of New Castle and the mouth of the Christiana, were six considerable streams, probably made so by the influx of the tide. These streams cut the ground up into necks or capes, the principal of which were named

"Grane Udden," or "Le Cap des Cruces;" "Strandwick," or "Baie de Rivage," and "Nieu Clareland." From New Castle to St. Georges, the land was known by the name of "Ackan Mamangaha Tusockhockung." The Appoquinimink was called the "Appoquonema;" the north point of Bombay Hook, "Hager Udden," or "Le Cap des Herons," (Cape of Herons); whilst Bombay Hook was called "Bomtes Udden." Duck creek was called "Aucke Kyhlen," or the "Reviere des Canards," or in other words, its present name anglicized into that of Duck river or creek. Jones creek was then called "Warge Kyghlen," or "La Riviere des Loups," translated to English, Wolfe river. Murderkill was named "Mordare Kÿlen," or "La Riviere des Assassins;" the name in English it at present bears. South of the Murderkill was called "Parades Udden," or "Le Cap de Parades," in English, Paradise Cape, promontory or point. From the above it will be seen that many places in our State still bear the old Swedish names, although many have been translated into English, and the idiom of some has been altered.

The Swedes that remained within the present boundaries of this State, generally settled near Fort Christina. Before the arrival of Penn, none of them resident within the limits of Delaware, lived more than a Swedish mile from the fort. A body of them clustered around Printz's house (afterwards occupied by his daughter, Armgardt Pappegoya,) and Fort Gottenburg, at Tinicum Island, where they built a small village called Printzdorp. The Fins, a distinct people from the Swedes, had their settlement in Pennsylvania, a few miles north of our boundary line, between Marcus Hook and Chester. The Swedes and Fins on the Delaware generally devoted themselves to husbandry, the Dutch to commerce. A great many of our most respectable families owe their descent to the Swedes, but their names have generally been anglicised. Amongst them the Rambos; the Justisons and Justises, whose original name was Gostafsson, the Boons whose name was formerly Bond, the Hoffmans whose names were formerly Hoppman, the Colesberrys changed from Kalsberg, the Wheelers from Wihler, the Yocums from Joccom, the Dalbows from Dahlbo, the Sinexes from Seneca, the Johnsons from Johansson, the Poulsons from Paulson, the Culins from Von Culen, the Vannemans from Van Neman, the Kings from Konigh, and the Keens from Kyn. The Walravens spelt their names with two l's. The Hendricksons are changed from Hindrickson, the Stidhams from Stedham. The Petersons, the Stalcops, the Matsons, the Talleys, the Andersons, and Walravens have their names spelt nearly as formerly, save in some cases the consonants are doubled, but the sound of the name is not changed by the spelling. The Vandevers, were spelt as Van der Weer, but their ancestors were Dutch, not Swedish, although they intermarried with the Swedes.

CHAPTER XIX.

FROM 1656 TO 1657.

Appointment of Derk Smidt as Governor—Departure of Stuyvessant—Indians assault Fort Cassimer—Jews on the Delaware—Jno. Paul Jaquette appointed Governor—The Council—Rules to govern the inhabitants—Swedes not to be allowed in the Fort—Vessels not to go above Fort Cassimer—Suspicion of the Swedes—Duties laid—Laying out of the town of New Amstel (New Castle)—Interview with Indians—Demand higher prices for their wares—Answer of Jaquette—Present given to them—Attempt to remove the Swedes into the villages—They remonstrate—First brick made in Delaware—Company object to written capitulation of Fort Cassimer—Swedish Ambassador demands restitution of New Sweden—Receives no satisfaction—Arrival of the Mercurius with 130 Swedes—Jaquette forbids her landing—Pappegoya petitions Council of New Amsterdam to let the Swedes settle on South river—They refuse—Mercurius runs past the batteries—Lands her passengers at Marcus Hook—Man-of-war Balance sent to South river—Mercurius sails for Manhattan—Is allowed to sail for Europe—Pappegoya leaves the country—His wife left at South river—Settlers from New Amsterdam arrive at South river—Build New Amstel (New Castle)—Lots granted—Territory between Christina and Bombay Hook transferred to city of Amsterdam—First bridge built in Delaware—Tobacco raised—Regulations—Inspectors appointed—Aldricks appointed Governor of Delaware—Waldenese expected—Stuyvessant ordered to purchase between North and South rivers—Fears New Amsterdam settlers will remove to South river—Fineness of the river.

After the conquering of the Swedes, the extending Dutch power over the Delaware, and the annexing our State to New Netherlands, Stuyvessant departed for the Manhattans, first appointing Captain

Derk Smidt as Commandant or Governor of Delaware. He was the first Dutch Governor of our State, and the sixth in succession from Minuit, the first Swedish Governor.

After the departure of Stuyvessant, late in the fall, Fort Cassimer was assaulted by more than five hundred Indians. The Dutch thought they were incited to this attack by the Swedes.[1]

At this early date we have an account of Jews on the Delaware, as in November, Abraham Lucenna, Salvador de Andrade and Jacob Cohen of that religion petitioned at Manhattan for the privilege of trading on South river. The petition was granted.

On the 29th of November, John Paul Jaquette (the ancestor of the celebrated Major Jaquette who fought so gallantly in the Delaware line in the Revolution,) was appointed the seventh Governor of Delaware.[2] He was to be assisted by a Council composed of Vice Director Andreas Hudde and Elmerhysen Klein. His instructions were as follows :

"If the affairs to be taken into consideration were purely military, or to relate to the company exclusively, they were to be assisted by two sergeants; but if purely civil, between freemen and the company's servants, then two freemen were to be chosen instead of the two sergeants.

"All meetings of the Council were to be called by him. He was to propose to them all matters relative to police, justice, and commerce. Things were to be

[1] Lambrechten, p. 109.

[2] Albany Records, vol. 10, p. 174.

decided by a majority of votes; the Governor was to have the casting vote.

"Hudde was to be the Secretary and Surveyor; he was to keep the minutes of the Council.

"They were strictly to observe the regulations respecting the sale of brandy, &c., to Indians, plundering gardens and plantations, running through the country, in town, and drinking on the Sabbath, its profanation, &c.

"No officers and soldiers were to be absent from the fort at night; no freemen, especially no Swede, living in the country to stay in the fort at night; Fort Cassimer not to be visited too often by them or the savages, much less the fortifications examined. In this respect, pay particular attention on arrival of any foreign vessel, yatchs or ships.

"No vessel to go above or below Fort Cassimer for trade with savages or Christians, but such as remain at the fort at anchor, and well near the shore. All on guard to be kept in good order, the fort to be kept in good repair, and its fortifications in a proper state of defence; but permission may be granted to plant on taking the oath to assist the fort, or be transported in case they refuse to take the oath.

"In granting lands, above all things, to take care that a community of at least sixteen or twenty persons reside together, or so many families, and to prevent coveting lands, require for the present, instead of one-tenth to be paid per morgan,[1] only twelve stivers per annum.

[1] About 7 acres.

"Grant no houses or lots on the side of the meadow of Fort Cassimer, viz.: between the creek and the fort, nor behind the fort, that land to be reserved for fortifying and outworks of the fort; for favoring the dwelling together on the south side of the fortress, lay out a convenient street behind the houses already erected, and lay out convenient lots on the same, about forty or fifty feet broad, by one hundred feet long, and the street at least four or five rood broad.

"Take very good notice of the behavior of the Swedes, and in case any of them should be found not well affected to the honorable company, and the state of our native country, to prevent further trouble, all such, with all imaginable civility, were to be caused to depart, and, if possible, sent to New Amsterdam."

Jaquette took the oath "to be loyal, and faithfully administer justice and maintain the laws, and maintain and protect the reformed religion, as learned and instructed, and in conformity with the Synod of Dordrecht, and to promote it as far as his power may extend, and to secure and defend the fort."[1]

This laying out of lots, &c., was the first commencement of the town of New Amstel, now New Castle. For a long time it was the most important town on the banks of the Delaware. New Castle is therefore the oldest town either in Delaware or Pennsylvania.

On the 20th of December, the following duties were imposed on liquors by Jaquette, viz.: On a a hogshead of French wine, 20 guilders; on an anker of brandy or distilled water per ton, 6 guilders; on

[1] Albany Records, vol. 10, pp. 174, 186, 191.

Holland or foreign beer per ton, 4 guilders. It was also forbidden to sell liquor to the Indians. Jaquette, previous to his administration on the Delaware, had for some years been a resident of Brazil.

On the 28th of December, several Indian sachems arrived at Fort Cassimer and demanded to be heard, which being granted in the presence of the Council and citizens. They stated that from the late commander (Delmetz ?[1]) they were promised an extension of the trade, and at higher prices. Jaquette replied, "that having but lately arrived, he did not know what was done before, but his wish was to live in peace and amity with them, and that if anything promised had been neglected through ignorance, it ought to be overlooked."

They asked an alteration in trade, "using a vast volubility of words," and demanded for two deer, a dress of cloth, or things in proportion. Jaquette replied, "that his principal's custom was not to dictate; but that each was at liberty to act his pleasure, and might go where his purse and the wares best suited," to which they assented.

They then stated that according to previous custom presents were made to the chiefs at the confirmation of a treaty. Jaquette answered "that goods were now very scarce, and though as much inclined as ever to give proofs of friendship, they would do however what they could." Accordingly, a subscription was

[1] Derk Smydt is undoubtedly meant. The records make sad work of names. Every prominent man's name has been spelt two or three different ways.

taken up amongst the traders of New Amstel, to which the company, and Jaquette also, subscribed liberally. Eighty-nine guilders was collected, with which presents were purchased and given to the chiefs. Two traders, Israel and Isaac Van der Zee refused to contribute, preferring "to depart the river and abandon the trade, rather than assist with the other inhabitants to maintain the peace and tranquility of the community."

[1656] Attempts appear to have been made in the beginning of this year to remove the Swedes by the Dutch, and settle them in villages according to the plan proposed previously, as on the 19th of January the "free people of the Swedish nation residing on the second corner above Fort Cassimer," appeared before the Council and solicited that they might remain on their lands, as they had no inclination to change their place of abode, or build in the new village, but claimed the promise made to them by Stuyvessant of being allowed to remain one year and six weeks, the time allowed by the capitulation. At the end of that time, they said they would conclude what to do.

At this time, it appears that bricks were made in this State, as "Jacobus Crabbe," on the 5th of February presented a petition to the Council "respecting a plantation near the corner, where *brick* and stone are made and baked." His petition was to be granted after the place was inspected.[1]

On the 13th of March, Stuyvessant received news

[1] Albany Records, vol. 10, p. 411.

from Europe, in answer to his despatch announcing the conquest of the Delaware. In reply they say, "We do not hesitate to approve of your expedition to the South river, and its happy termination. While it agrees in substance with our orders, however, we should not have been displeased that such a formal capitulation of the surrender of the forts had not taken place, but that the whole business had been transacted in a similar manner, as the Swedes set us an example of when they had made themselves master of Fort Cassimer; our reason is, *that all which is written and copied is too long preserved, and may sometimes, when it is neither desired or expected be brought forward, whereas words not recorded, are by length of time forgotten, or may be explained, construed or exercised as circumstances may require.* But as all this is passed by, our only object in making this observation, is to give a warning if similar opportunities might present themselves in future. You will take care that said Fort Cassimer is in every respect well provided, and placed in a state of defence, but do not mind the Fort Christina, leaving only to ascertain its possession, three or four men in it, soliciting some individuals to establish themselves there."[1]

The news of the conquest of New Sweden had by this time reached the Swedes. Accordingly on the 22d of March, H. Appleboom, the Swedish resident at the Hague, delivered a strong protest to the Dutch government respecting it. In it he asserted that the Swedish company had the best title to the territory;

[1] Albany Records, vol. 4, p. 204.

that they had purchased it from the natives. He desired speedy redress, and that the Swedish company should be restored undamaged. The States General received the memorial, and passed a resolution on the 24th of March, that it should be examined, together with an extensive memorial on the "same business by the deputies of Amsterdam, Rotterdam and Hoorn," and a speedy answer made to the Swedes.[1] But nothing was ever done to give satisfaction to the Swedes, and it continued to be a subject of complaint on the part of the crown of Sweeden to the State General for many years. In this correspondence the Delaware is called "the South river of Florida."

Before the news of the conquest of their possessions on the Delaware had reached Sweden, they had despatched a vessel from there called the Mercurius, with 130 people for South river. She arrived about the latter end of March. Jaquette prohibited her landing her crew and passengers, notwithstanding the request of Pappegoya and Swen Schute, who remained after the departure of Rising, and who were no doubt glad to see the arrival of friends and countrymen from fatherland. Word was immediately sent to Fort Amsterdam. The Council there met on the night of the 28th of March, and passed a resolution confirming the action of Jaquette in prohibiting the landing of her crew and passengers, but allowing her to revictual and depart for Sweden. Two days after the passage of this resolu-

[1] MSS. in N. Y. Hist. Soc.

tion a letter was received by the Council at New Amsterdam from Pappegoya requesting that those that came from Sweden should be allowed to settle in this country. Hendrick Huygen, the captain, also wrote to the Council, making the same request. In his letter complaining of the hardship that would be the result if the colonists in his vessel were not allowed to land, he said, "beside the immense loss they would suffer, many good farmers would be ruined, parents separated from children, and even husband from wife." From this it appears that many that came over in the Mercurius were portions of the families of those already resident on the Delaware. The Dutch desired them to settle at New Amsterdam. But the captain, Huygen, very properly remonstrated against his countrymen "being compelled to reside among a foreign nation whose language they could not understand, and whose manners were unknown to them." This remonstrance also had no effect upon the New Amsterdam Council, who at a meeting on the 11th of April more peremptorily insisted that the crew and passengers of the Mercurius should not land on the South river.

The Dutch were greatly alarmed at this accession to the strength of the Swedes, and it was resolved at the same Council to send the man-of-war Balance to the South river to bring the Mercurius to Manhattan. But the Captain arriving overland to hold personal communication with the Council, to save expense, this part of their resolution was rescinded, although it was resolved to keep her at New Amsterdam until the Mercurius arrived there. In the meantime rumors came to them through the savages that there had been difficulty between the Swedes and Dutch. On the strength of these rumors, twelve or fifteen armed men, under command of Ensign Smith, were at once sent to the South river overland. Information, however, soon afterwards arrived that John Pappegoya and a number of Indians and resident Swedes had gone on board the Mercurius and sailed past Fort Cassimer, both in defiance of Dutch orders and Dutch cannon, and landed her cargo and passengers at Marcus Hook. The Dutch, it is said, did not like to fire on the ship for fear of injuring the Indians. In the Mercurius came over a Swedish clergyman named Matthias. He stayed here about two years, and then departed for Europe.

The Dutch were much incensed at this action of the Swedes, and despatched the man-of-war Balance, whose previous order for sailing to the South river they had rescinded, to bring the Mercurius to New Amsterdam. This was done, but it appearing upon examination that the Captain had nothing to do with running past the Dutch Fort, but that Pappegoya and the resident Swedes were the only ones to blame in the transaction, he was allowed to return to Europe with his vessel upon paying the duties on the cargo. Pappegoya about this time must have returned to Sweden, leaving his wife, Armigard, the daughter of Governor Printz, behind him, as on the 3d of August, in her own name, she petitioned the Council at New Amsterdam to be allowed to take possession of

Printzdorp and Tinnakonk, which she had been forbidden to do by Jaquette. Her petition was granted. Printzdorp, it is supposed, was situated on Tinicum Island. It is more than probable the reason for Pappegoya's leaving the Delaware was the part he took in running the Mercurius past the batteries. It is believed he never returned to Delaware, as no mention is ever made of him afterwards in any record in relation to the affairs of this State. Frequent mention, however, is made of his wife Armigard.

In the spring a number of settlers came from Fort Amsterdam and settled at Fort Cassimer. The Governor and Council then commenced giving deeds for lands. By August they had given seventy-five, mostly for lots in the town of New Amstel, around the fort. A yearly rent of twelve stivers for every morgan of land was required.[1] Within the Swedish districts, which were principally on both sides of the Christina and Brandywine creeks, in New Castle, Brandywine, and Christiana hundreds, in New Castle county, no deeds were given at this time, but a tax of five or six gilling was laid on every family agreeable to the project of the Schoute.[2]

From various causes, the Dutch West India Company now found themselves much in debt. To recover the South river from the Swedes, they had to receive aid from the city of Amsterdam. To relieve themselves from this embarrassment they on the 16th of August agreed to transfer to that city all the lands

[1] Acrelius. A stiver is a Dutch coin, in value about two cents.

[2] Acrelius, p. 420.

on the west side of the Delaware, from the south side of Christina creek[2] to Bombay Hook. This was called the Colony of the City. All the land above Fort Christina, extending up the river as far as the limits of the Dutch settlement was called the Colony of the Company. Acrelius has the matter exactly contrary, giving the Colony of the Company jurisdiction below the Christiana, and that of the city above, or north of it; and Ferris, in his "Original Settlements," follows his authority. But a copy of the original grant is extant, and republished in the Holland Document. The transfer, however, was not concluded until April 12, 1657. The following was the agreement entered into between the Burgomasters of Amsterdam and the intended settlers of the Colony of New Amstel, (now New Castle,) that they were to transport from that city to the Delaware.

They were to be transported, with their families, furniture, &c., in vessels to be procured by the city, who were to advance the freight money, to be afterwards refunded. The city engaged to provide them a fruitful soil, healthy and temperate climate, watered by and situated on a fresh water river, on which large ships may sail, having made an agreement with the West India Company to this effect, for a place at their disposal, to which no other persons have any claims. The city was there to lay out on the river side a suitable place for their residence, and fortify it with a trench without and a wall within, and divide the inclosed lands into streets and lots for traders, mechan-

[1] The Indians also called Christina creek, Suspencough.

ics and farmers, and have a market, all to be at the expense of the city. The city was to find a schoolmaster, and provide for him, who should also read the Holy Scriptures in public, and set the psalms.

The city was also to provide for one year, clothing, provisions and garden seeds; to build a large storehouse to contain their goods, clothing and provisions; to keep a factor there with all necessaries, and sell them at Amsterdam prices to the colonists. The company's toll was to be employed in building and supporting public works, under direction of those authorized by the West India Company and city. The police of the town and city, as well as administration of justice was to be as in Amsterdam. A schout, or head justice was to be appointed in the names of their "High Mightinesses and West India Company," by deputies of Amsterdam, who were to give the director a power of attorney. Three burgomasters were to be appointed by the common burghers, from the "honestest, fittest, and richest" of the inhabitants, and five or seven schepens, of which the burghers were to nominate a double number, from which the director, by attorney, were to select. When the town contained two hundred or more families, they were to choose a common council of twenty-one persons, to meet with the burgomasters and schepens, and consult on matters relating to government of the city; once established, the council were to supply vacancies by a new election, by a plurality of votes. They were annually to choose burgomasters, and nominate a double number, out of which to choose schepens. The schepens were to decide causes for all sums under *one* hundred guilders[1] ($60); over that sum appeal was to be allowed to the Director General and Council of New Netherlands. The schepens to pronounce sentences in criminal cases, subject to appeal.

The city of Amsterdam was to agree with a smith, wheelwright, and carpenter to live with the colonists.

The city agreed to divide the lands about the town into fields, for plough, meadow and pasture, and allow for roads. Every farmer was to have "*in free, fast and durable property*," as many morgans of land as the family could improve, and for grazing, which was to be under cultivation in two years, or they were to be forfeited. No poundage, horn or salt money was to be required for ten years from the first sowing or pasturing; at the end of ten years they were to pay no other tax than the lowest paid in any other district of the West India Company. They were to be free from tenths for twenty years, from the first sowing; at the end of twenty years the tenths were to be given to the city of Amsterdam, half of which tenths were to be applied to the support of the public works and of persons employed in the public service, so also of poundage and other charges whenever paid.

The ships sent from Holland by the city of Amsterdam were to load and bring over corn, seeds, merchandize and wood. The colonists were at liberty to charter private ships; but they were to be consigned to the city of Amsterdam, who were to provide

[1] A guilder is about 60 cents.

for them storehouses and sell the goods, and return the proceeds as ordered, deducting only two per cent. for commissions, and one tenth of net proceeds for disbursements by the city for the freight and passages of the colonists, and their goods, which were to cease when the disbursements were refunded.

The colonists were allowed to take what they required from the city's warehouse, at a fixed price, the account to be sent with their goods and to be deducted therefrom.

The colonists were allowed to cut what wood they require for building houses and vessels, from any forest in the district not granted to private persons, free of cost, and were allowed also freely to hunt in the woods, and fish in any waters not private property, under regulations to be made by the company, &c. The burgomasters of Amsterdam, as founders and patrons, were to appoint secretaries, messengers, and other inferior officers, and the city of Amsterdam were to see that all tools and implements were transported free and without recognitions. The discovery of minerals, crystals, precious stones, marbles, &c., were allowed to keep them as their own for ten years, free of duty or imposition; after ten years they were to pay to the company one-tenth of their proceeds. The city of Amsterdam was to provide a warehouse for all the goods it should send to New Netherlands, to be visited by any person appointed by the directors of the West India Company, under the inspection of a person appointed by the city, and marked with the marks of the city and company; the recognition was to be paid to the company, according to the list. These goods might then be laden with the knowledge of the company, in any vessel they could obtain. If the city was to ship any goods or freight to New Amsterdam, they were to submit to the same regulations as others; but if the city of Amsterdam sent its own or chartered vessels, laden only with its own goods, the vessel was to go to its own city or colony, but all the goods on board were to be advertised in the city warehouse, under the inspection of any one of the company appointed for that purpose, to whom the letters and commission from the company were to be delivered; and *vice versa* of goods that were imported from the city's colony into Amsterdam. Duties to the country and to the company were to be paid.

All materials, &c., for farming or trades were to be free from recognitions; all produce of New Netherlands on importation was to be free of duty; so also was salted or dried fish taken there. Peltry, such as beavers, otter, &c., had to pay eight per cent. Besides the duties above paid in Amsterdam, there was also to be paid in New Netherlands 4 per cent. in light money, reckoning the rix dollar at 63 stivers.[1]

On the 4th of November, a bridge was ordered to be built over the creek near Fort Cassimer. This was the first bridge that we have any account of as being built within the limits of this State.

[1] Hazard's Historical Collections, vol. 2, pp. 543, 547.

It appears that tobacco was at this time one of the most important productions raised by the Dutch in this State. On the 9th of November the inhabitants of New Amstel were summoned to nominate four persons, out of whom to elect *two expert* persons as tobacco inspectors.

On the 12th of December a placard was posted up stating "that it is ascertained *that this river has become renowned for its tobacco,* and persons are daily encouraged to settle within its limits; if this produce is managed as it ought to be, and properly cured and packed, and all sorts of fraud, which might be used, is carefully avoided; and whereas, all this cannot well be effected except it is examined and inspected before reported, and whereas expert and faithful persons are required for this inspection, &c. Meenes Andriessen and William Maunts are selected for the nomination as inspectors." They were to inspect all the tobacco before it was exported. Jaquette commanded that no tobacco should either be delivered, received or exported without being inspected. For each hundred weight of tobacco, six stivers were to be paid by the receiver, and four stivers by the seller, under penalty of fifty guilders for the first offence, and for the second and third in proportion. The informer was to receive twenty guilders, and his name was to be kept secret.[1]

In the latter part of this year, Stuyvessant received information of the sale of Fort Cassimer and the territory in our own State, from Christina to

[1] Albany Records, vol. 10, pp. 448, 450.

Bombay Hook, to the city of Amsterdam. The name of Fort Christina was then changed to *Altona.* He was also informed that Jacob Alricks (now spelled Alrich,) was appointed Director or Governor of the City's Colony. The numerous and respectable family of Alrich in this State are descended from the nephew of this Governor. One of them now occupies the land which has been successively inherited in unbroken succession from his ancestor, this nephew, who was also a governor of Delaware. Stuyvessan't was instructed to garrison the different forts of Altona, New Gottenberg and the Island of Hattenberg, with eight or ten soldiers. The directors also informed him that "to all appearance a large number of the *exiled Waldeneses*" would flock to Delaware as an asylum. They promised to endeavor to induce them to do so. They also sent him orders "to endeavor to purchase, before it could be accomplished by any other nation, all that tract of land situated between the South river and the corner of the North river." The directors also express their fears to him that the establishing of the new colony at Delaware, and the consequent cheapness of goods and taxes may prove an injury to New Amsterdam, by the migration of its citizens there. In a Holland document, written this year, reference is made to the Delaware as being considered "the finest river of all North America, being wide, deep, and navigable; abounding in fish, especially an abundance of sturgeons, of whose roes a great quantity of cavejaar might be made."[1]

[1] Holl. Documents, vol. 8, pp. 32, 50.

CHAPTER XX.

FROM 1657 TO 1659.

Meeting at New Amstel to regulate the price of furs—The penalty for its violation—Trading with Indians forbidden—The city of Amsterdam sends settlers to Delaware—Wreck of the Prince Maurice—Arrival of Alrick Governor of New Amstel, of D'Hinoyossa, of Martin Krygier—Formal cession of New Amstel to city of Amsterdam—Deed of grant—Arrival of Alricks at New Amstel, settlers and soldiers—Van Sweringen—Removal of Jaquette—Twelve Indian Nations.—Attempt to get the Swedes to settle in villages—They have their own officers—Murder of Laurens Hansen by Indians—City Hall and 100 houses built at New Amstel—Its government—Seven City Councillors elected—Schepens appointed—Salt Works on Delaware—Bricks at New Amstel.—Sickness—Alricks uses the company's negroes and oxen—Cows purchased—English wrecked at Cape Henlopen—Ransomed—Settle in Delaware—Jealousy of the West India Company—No more English to settle—First Church and Schoolmaster—Forts New Amstel and Altona—Food scarce—Bad state of affairs—Council at New Amsterdam on affairs of Delaware—Frauds of merchants and inhabitants—Stuyvessant visits Delaware—Swedes take the oath of allegiance—Privileges granted them—Subsidies from the Swedes to be neutral in war between Sweden and Holland—Jealousy of English—Stuyvessant instructed to purchase land between Bombay Hook and Cape Henlopen—Sickness at New Amstel—Distress for bread—Arrival of ship Mill with settlers—Sickness amongst them—Beekman, Governor of Altona—His instructions—First tailors, shoemakers—Expenses of garrison at New Amstel.

On the 10th of January, a meeting was held at New Amstel to fix the price to be given to the Indians for their skins and furs. Complaints were made that some persons did not hesitate [1657] to spoil the trade with the Indians by increasing the price of deer skins one-third, which price it was feared would continue to be increased to the "great loss of the lower classes." Those "who made their living by their hands, it was stated, would have to pay a higher price for several articles than they could sell them for again." This inconvenience was thought to be small, compared to "what might be apprehended when the spring trade in beavers was opened. In which case the inhabitants residing in the neighborhood might be utterly ruined." Jaquette, who called this meeting, stated that "serious complaints had been made to him in relation to the matter, but he saw no mode to avert the evil, save by calling a meeting of the community and determining on a price which should be adopted as a rule in trading. He promised that all orders the community should decree should be promptly executed by him and the Council." Accordingly they adopted the following rules, "and promised by their signatures, on their honor and oath" to maintain them. If they broke them, they desired to be considered perjured men. For the first violation, the penalty was to "forfeit the privilege of trading for a whole year." For the second such penalty as the community might inflict. For the third violation they were to be expelled from the river. The prices agreed upon were as follows:

For a merchantable beaver, two fathoms of seawan.

For a good bear's hide to the value of a beaver, two fathom.

For an elant's (moose) hide the value of a beaver.

For a deer skin 120 seawan.

Those of foxes, ratelaplan, hispans, and others in proportion. This was signed by about thirty-six persons, of whom about a dozen made their marks.[1]

On the 10th of January Jaquette issued a placard complaining of the great abuses by the inhabitants running after the savages, and detaining them, when they have articles for sale, and orders "that in future no person shall go to the Indians by land or water to trade with them, or offer them by gifts by sailing up and down the river; nor meet the Indians when they approach their dwellings to hire and coax them, or call them in their houses, but permit them to use their pleasure," under penalty of the forfeiture of the furs thus obtained, besides arbitrary correction.

In the meantime the city of Amsterdam prepared to send out settlers to their colony in Delaware. They fitted out four vessels, which sailed from Amsterdam the latter part of 1656 or beginning of 1657, and arrived on our coast in the early part of March. They were the ship Prince Maurice, which had one hundred and twelve persons on board, besides a crew of sixteen officers and sailors; the Bear, which had thirty-three persons, and the Flower of Gelder, which had eleven, being in all one hundred and sixty-seven souls, who intended to settle in this State. There was also a vessel called the Beaver, the number of whose passengers is not given. They all arrived safely but the

[1] Albany Records, vol. 10, p. 455.

principal ship, the Prince Maurice, which was wrecked a few days previous to the 12th of March, on Long Island, near the present town of Islip, near Fire Island Inlet. The crew and passengers were saved. In the Prince Maurice came the new Governor for the City's Colony, Jacob Alricks, and a company of fifty soldiers, commanded by Captain Martin Krygier, and Lieutenant D'Hinoyossa. The ship proved a total loss, although a great portion of the cargo was saved. D'Hinoyossa afterwards became Governor of the State, and Martin Krygier held several important offices, and remained here after the conquest of the State by the English. Alricks wrote an account of the disaster to Stuyvessant, who promptly sent him relief.

In addition to the former mentioned vessels, the city of Amsterdam also sent over a few months later, the man-of-war Balance, (which by this time had arrived in Amsterdam), with colonists, and a sloop called the Golden Mill, with merchandize.

On the 12th of April, Stuyvessant made a formal session of the lands on the Delaware, heretofore mentioned as purchased by the city of Amsterdam, to Jacob Alricks,[1] who immediately afterwards em-

[1] The following is the grant:

"I, Peter Stuyvessant, in behalf of the high and mighty Lords, the States General of New Netherlands, and Lords Directors of the privileged West India Company, Director General of New Netherlands, &c., declare that on this day, in conformity with the orders and letters of the Directors, dated December 19, 1656, I made a transfer and cession to the Honorable Jacob Alricks and Commissary General of their Colony on South river, in New Netherlands, the *Fortress Cassimer*, now named *New Amstel*, with all the lands dependant on

barked on the Gilded Beaver for New Amstel, the future seat of his government. He took with him one hundred and eighty people, of which seventy-six were women and children, and sixty soldiers. They reached New Amstel after a passage of five days, about the middle of April, when he assumed the government of that portion of Delaware from the southerly side of the Christiana, to what is now known as Little Duck creek. Jaquette remained governor of that portion of the State on the northern side of the Christina, comprising the city of Wilmington and

it, in conformity with our first purchase from, and transfer by the natives to us, on the 19th of July, 1651, *beginning at the west side of the Minquas, or Christina Kill,* (named in their language Suspencough), *to the mouth of the bay or river included, named Bompjeshock,* (Tree's Corner), in the Indian language Canarasse, and this so far in the country as the limits of the Minquas land, with all the streams, hills, creeks, harbors, bays and plains belonging to these; of all which lands, with their appendages and dependencies, we now make a cession and transfer in the name and behalf of the Lords Directors, patrons to the aforesaid, the Hon. Jacob Alricks, in behalf of the Honorable Burgomasters and rulers of the city of Amsterdam, making a cession of all our actual and real possessions, property, right and privileges, and all that on such conditions as between the aforesaid Directors, and the Burgomasters and rulers of the city of Amsterdam, have been sanctioned; appointing, therefore, in our place, and constituting the aforesaid Hon. Jacob Alricks in that quality, in behalf as before proprietor, in our place, without reserving to our place in our former quality any action or pretension, promising, therefore, to hold sacred this our transfer. In truth whereof we have signed this, and confirmed it with our usual signature.

"Done in Fortress Amsterdam, New Netherlands, April 12, 1657.

P. Stuyvessant,
Nicolas De Sille,
Peter Tanneman."

Albany Records, vol. 15, pp. 124, 125.

the hundreds of Brandywine, Christiana, White Clay creek, and Mill creek, together with the portions of Pennsylvania that were then settled. Gerritt Van Sweringen, afterwards one of the governors of this State, was the supercargo of the Gilded Beaver.

Jaquette held his office as governor of the northern portions of Delaware but a few days after the arrival of Alricks, for complaints having been made to Stuyvessant of his "delaying and declining to administer justice, obstructing legal arrests, of arbitrary executions on his own authority, without the shadow of any legal process, but by acts of violence, of taking possession of lands, and cultivating those which were granted to other persons;" for these, and various other alleged breaches of law and right, Stuyvessant, on the 20th of April, removed him from office, and instructed him to deliver all the effects of the company into the hands of Andreas Hudde, Jan Juriansen and Sergeant Paulus Jansen. He was also instructed to prepare for his defence. He at the same time sent Captain Martin Krygier overland to Delaware, to receive what goods Jaquette had had in his possession. On the 23d of May, Attorney General Nicatius De Sille arrested Jaquette. In asking authority for this course he asserts that Jaquette "vexed the community, tyrannized over the inhabitants, and made the soldiers' lot nearly insufferable." Great complaints were also made by the farmers against him. On the 18th of June Jaquette appeared before the Council, where the main body of these charges do not appear to have been inquired into, but instead, some disputes with Jacob Swenske, as to "who violated the arrest of a certain Neil Swarsen," received the attention of the Council. As to the result of the trial of the seventh governor of Delaware, history makes no mention.

Alricks soon after his arrival made inquiries in relation to the number of the Indian nations in his vicinity—he found there were twelve of them.

Attempts were again made to get the Swedes to settle in villages. They appear at this time to have had their own officers. Gregorious Vandyke was their sheriff.

On the 10th of August, Laurens Hansen, described as a captain of armes, one of the garrison at Altona, (now Wilmington), went with Alexander Boyer on a trading expedition to the Minquas, and was murdered by the savages, and robbed of some seawan, and a few other articles which he had with him. The Minquas chief, who resided in the next fort to Altona, recovered some of the articles from the murderer and restored them to the Dutch. It does not appear that any punishment was ever inflicted on the Indian murderers (whoever they were) for this offence.[1] This was the first Indian murder in New Castle county, and with the exception of the massacre at Lewistown, the first murder known to have been committed in the State.

In the meantime the building and improvement of New Amstel went on energetically. About the beginning of September, lots were granted to the colonists. A magazine was erected, the fort repaired, a

[1] Albany Records, vol. 12, pp. 424, 425.

guard-house, bake-house and forge built, together with residences for the clergymen and other public officers. A city hall, for the burghers, was also erected. It was a log building, two stories high, and twenty feet square. The whole of the public buildings were enclosed within a square. At the end of the first year New Amstel was a handsome little town of about one hundred houses.[1] It was therefore nearly one-half its present size. Alricks, in a letter written about this period, thus speaks of the government of New Amstel. He says:

"I found the government to consist of a military council over the soldiers, who were here of old. The differences between the old settlers, who consisted of about twelve or thirteen families, were decided by the commander and two persons acting as schepens, and a secretary appointed from among the inhabitants, by the general, on the part of the West India Company. These expressed a desire now that the place had changed hands, that a burgherlike government should be continued, according to the conditions, as it was under the Director General and the West India Company. So it was that they continued to decide all differences between burgher and burgher. All affairs appertaining to the city and military matters were disposed of by me and the Council, and differences between the city's servant, soldiers and trainsbands and freemen, until the arrival of the Balance, when seven city councillors were elected, and from them three new schepens were chosen; an-

[1] Holl. Documents, vol. 15, pp. 12, 213, 225, 227, etc.

other secretary and schout were also appointed, two elders, and two deacons for the management of church affairs."[1]

Salt works are referred to in the records at this period. Bricks for Delaware appear to have been brought from Fort Orange, now Albany. Things did not appear to go on well at New Amstel either. On the 15th of September, complaints were made that Alricks used the company's oxen and negroes. In the latter end of October, there was also a great deal of bilious fever at New Amstel. Alrick's wife and three or four children were sick. Forty cows were at the same time introduced in the colony, which were purchased by Alricks at prices ranging from 128 to 130 guilders per head.[2] This would make the price of a cow at that time about $78 80.

About this time two boats, with fourteen Englishmen ran on shore at Haverkill.[3] The Dutch despatched a vessel to save them, but did not succeed in their attempt, though they lost an anchor. They however managed to ransom six of them from the savages, and brought them to New Amstel. Alricks immediately sent information to the Governor of Virginia, as he supposed[4] they belonged to that place. They however settled at New Amstel. When the news of this affair reached Amsterdam, it caused a great deal of uneasiness to the Dutch. They were afraid that the English were endeavoring to get pos-

[1] Holl. Doc. vol. 2, p. 337. [2] Ibid, p. 437.

[3] This place appears to have been near Cape Henlopen.

[4] Albany Records, vol. 12, p. 437.

session of the country. They accordingly wrote to Stuyvessant to instruct Alricks if these men were fugitives from labor from Virginia, he was to return them. If they were freemen, he was to get rid of them the best way he could, without giving offence, but on no account to let any more English settle there, "much less to allure them by any means whatever." From subsequent accounts, it appeared that the whole fourteen were ransomed from the savages, and that all finally settled in this State.

At the latter end of this year there appears to have been further troubles at New Amstel. At any rate, Alrick's letters from there were filled with complaints. Several residents of New Amstel who had purchased goods of the company, and mortgaged their houses and lots as security, sold them, and moved to Altona without satisfying the company's debt. On the 15th of December, the company issued a warning to purchasers that such sales were null and void, and cautioned them not to make payment unless with the company's consent.[1] Fort Cassimer he represented as in sad condition; the expenses heavy, the means gone, no magazine to save victuals, the walls and fortifications crumbled down so that it was as easy to pass them through the walls as through the gate. Another gate, he said, was required, to make some outward appearance of defence. Some parts of the fort had been washed away by encroachments of the river. He also represents Fort Altona (Christina) as decaying, and stated that it had had no garrison for

[1] Albany Records, vol. 12, p. 437.

a long time. He complained of the scarcity of food for the garrison, and of his want of means to get any. Complaint had also been sent to Stuyvessant in relation to his administration by Ensign Smith and Henry Huygen,[1] which gave him additional grounds of discontent.

The Delaware was about this time frozen over opposite New Amstel, in one night, so that deer could run over on it, which the Indians relate had not happened before in the memory of man.[2]

On the 25th of April, Evert Pieterson, whose official position was that of schoolmaster and comforter of the sick, landed at New Amstel. He is the first schoolmaster of whom there is any record on the Delaware. He at once commenced keeping school, and had 25 scholars on the 10th of August following. In a letter of his to the Commissioners of Amsterdam, he states that "wharves were already laid out" at New Amstel, "and almost built." He also says that he "found 20 families, mostly Swedes," in the City's Colony, (that portion of Delaware south of the Christina), "and not more than five or six belonging to our (the Dutch) nation."[3]

[1658] In the spring things were still backward in New Amstel on account of the prevalence of the bilious fever in the fall. Building was slow, as from the desponding letter of Alricks giving an account of affairs on the South river, we learn that

[1] Alrick's letters, in Albany Records, vol. 4, p. 283.
[2] Canpanius, p. 55.
[3] Broadhead and O'Callighan, vol. 1, p. 7.

there were only "four or five carpenters" at New Amstel, "and they were usually poor workmen, without experience or ingenuity." They were also short of provision, but a few heads of cattle arrived from Virginia, and that somewhat revived them.[1]

On the 20th of April, the Council met at New Amsterdam, and took into consideration the affairs of South river. It was decided that there must be a change of management, as owing to the removal of Jaquette there were many irregularities. Stuyvessant stated he was advised of great frauds by merchants of this city, (New Amsterdam), and others on South river, in non payment of imports and exports. "That those who did not pay, could sell cheaper than those who did." That several persons of New Amstel asked permission to settle near Altona, "there to begin plantations." Regulations, it was said, were necessary among the Swedes. It was decided that Stuyvessant and Peter Tonneman should proceed to the South river.[2] This they accordingly did. They were met at Tinicum by the officers of the Swedes, viz., Gregorious Van Dyck, the Sheriff, Oloff Style, Mathys Hanson, Peter Rambo, and Peter Cock, Magistrates; Swen Schute, Captain; Andrier D'Albo, Lieutenant, and Jacob Swensen, Ensign. They renewed their allegiance, and presented a petition for the following measures, viz.:

For executions they asked a Court Messenger.

Stuyvessant told them "the jailor could perform this duty, as he was then employed by the sheriff

[1] Alrick's letter, in Albany Records, vol. 4, p. 285. [2] Ibid.

and commissioners to make summons, arrests and executions."

They asked for "free access to the soldiers at Altona, in case they wished their aid for execution of resolves."

Stuyvessant commanded "the provisional commissary to furnish them if solicited by the sheriff."

They asked that "no person should leave their limits without the knowledge of the magistrates, much less male and female servants, that when they leave without a discharge, or try to run off, they may be arrested."

Stuyvessant ordered "that no person should leave without the consent of the commissary. Consent, however, was first to be obtained of the Director General and Council, as usual in New Netherlands; and if any Swedes were to depart, the sheriff was to order him to return, and in case of refusal, to arrest him, and inform the Director."

Some subsidies being required, the sheriff and commissary, were directed to inquire "where they could be obtained with the least incumberance to the Swedish nation," who are styled "our good and faithful subjects, whom we promise that we cordially desire to favor as much as any of our own nation."

The Swedes after taking the oath of allegiance, demanded that if there arose a difference between Sweden and Holland in Europe, that "they might be allowed to remain neutral, and side with neither party." This request was granted by Stuyvessant.

The Directors at Amsterdam were still haunted

with visions of English encroachments on the South river, and fearing they might endeavor to purchase the lands in the State between Bombay and Cape Henlopen, then called the Hoernkill. On the 1st of June, they sent instructions to purchase it from the Indians, on the account of the Colony of the City, who pledged themselves to erect a redoubt for its defence. They further informed Stuyvessant they intended to place buoys in the bay for the security of vessels which might arrive on the coast.

In the fall of this year, New Amstel was again badly afflicted with the "fall" or "bilious fever," and to add to the calamity, the barber (surgeon) died. But few old people died, the mortality being chiefly among the children. Amongst the sick were Hinoyossa, and Rynvelt, the commissary, who afterwards died, and all the schepens. Christian Barents, whom he had employed to build a Ross Mill, also died. The number carried off by this sickness amounted to about one hundred, or one-sixth of the population. As the population of the Colony of the City was then according to a letter of Alricks to the Commissioners, appointed by the City of Amsterdam, six hundred souls. The colony was also in great distress for the want of bread and corn. The harvest proved a failure. The worm appeared in vast quantities, and injured the crops and gardens. It also suffered from drought, and then again from excessive rains. It did not even produce its seed, as where nine hundred sheples were sown, only six hundred sheples were produced. Rye was sold at nine guilders the sack; peas at seven or eight guilders per gallon. Again there was not a single merchant in New Amstel that sold provisions. The feeding for one year, also from the company's stores, had caused many of the people to neglect work. Alricks, in one of his letters, says: "Many come here poor as worms, and lazy with all, and will not work unless compelled by necessity."

Several children were sent at this time from the Almshouse of the City of Amsterdam to New Amstel. They were bound out by Alricks for two and three years.

On the 27th of September, the ship Mill arrived from Amsterdam, with one hundred and eight settlers on board. Owing to the long voyage, scurvy broke out amongst them, and ten of her passengers died. Three more died after their arrival. She brought no provisions, and so many more mouths being added to consume the slight stock of provisions on hand, caused additional distress.

Notwithstanding the general distress, Alricks greatly improved the town of New Amstel by the erection of several buildings and enlargement of others. He built a barracks adjoining the fort, of 119 feet long, by 16 or 17 feet wide; a public store of 27 feet long, and 54 feet wide; a bake-house roofed with tiles imported from Holland, 18 feet wide, and 31 or 32 feet long. This house was built in the Square, (more than probable the square where the public buildings stand in New Castle), and a house for the commissary. He also purchased and enlarged a building to be used for a church, also a house for the minister. This is the first mention made of any building used solely for religious purposes in this State.[1] We have no evidence in regard to the place in which it was situated. The erection of the barracks were for the soldiers who had wives. Most of the soldiers were married, and had servant girls, and drew rations for themselves, their wives, and servants, from the company. The position, therefore, of a soldier, must have been better in those early days than now. Many of the settlers, Alricks complains, were "weavers, shoemakers, buttonmakers, and tailors. Farming," he alleges, "was too hard for them. They did no work, but loaf about."[2]

On the 28th of October, William Beekman, a Schepen and Elder of New Amsterdam, was appointed Vice Director and Governor of Altona, in place of Jaquette. He was to be supreme commander in that part of our State, both in civil and military affrirs. His salary was the same as Jaquette's, viz., 50 guilders per month, and 200 guilders per annum for board, in all 480 guilders per year. His residence was for the present to be in the dwelling house at Fortress Altona, but he was instructed, as soon as possible, to have his permanent residence at or near New Amstel, and to hire a house or rooms for that purpose at the expense of the company.

[1] This was undoubtedly the commencement of Emanuel Church at New Castle.

[2] Albany Records, vol. 12, pp. 467, 476. Broadhead, vol. 1, pp. 49, 56.

Delaware was therefore at this time divided into two States, with two governors.

He was, amongst other matters, instructed on the arrival of any vessels or yachts of any nation (or at least before their unloading), to be in or near Fort New Amstel, to attend carefully to their loading and unloading. To allow no goods to be laden or unladen without his examination, and to see that all duties were paid. To prevent smuggling, he was always to have a guard of the company at New Amstel, under his orders. He was to seize all smuggled goods, and have a share of those confiscated, and prosecute the smugglers before the Council. From their decision there was an appeal to New Amsterdam. He was also to have all the powers possessed by the company in Altona, to administer justice both in civil and military affairs, and in criminal cases of minor grade. He was also instructed to find out the owners of land between Bombay Hook and Cape Henlopen, if their demands were reasonable, to enter into an agreement for their purchase. He was to take the advice of Alricks in his purchase, and if he had an opportunity before winter, to erect a fortification at Henlopen or the Hoernkiln.[1]

The wages for labor at this time in Delaware, according to Alrick's letters, were for laborers three guilders a day, for mechanics four guilders a day.

The estimated expenses of the garrison of Fort

[1] Beekman's Letters, which have been preserved amongst the Records at Albany. They form the most valuable history of the early settlement of this State by the Dutch.

Cassimer were as follows: Captain, 50 florins per month; lieutenant, 30; ensign, 25; two sergeants, 30 florins; one captain of arms, 10 florins; two corporals, (12 florins each), 24 florins; six cadets, (each 10 florins), 60 florins; two drummers, (9 florins each), 18 florins; forty-four soldiers, (each 8 florins), 352 florins. Total pay per month, 599 florins. The expense in addition for rations was for the captain for the year, 150; the lieutenant, 120; the ensign, 100; each sergeant, 80; and each soldier 60 florins. The estimated expense of the garrison of New Amstel was 11,018 florins per annum.

CHAPTER XXI.

A. D. 1659.

Ravages of fever at New Amstel—Death of Alrick's wife—Desponding letter of Alrick—West India Company suspicious of the Swedes—Disapproves of their arming and appointing their officers—Wish them settled among the Dutch—Alterations of the agreement with settlers emigrating—Consternation and dismay of the Colonists thereat—Emigration to Maryland and Virginia—Sickness and bad harvests at New Amstel—Scarcity of food—Provisions shipped for New Amstel ran away with—Deaths among the citizens—Purchase of land between Bombay Hook and Hoernkiln—Swedish minister forbid to preach—Descriptions of settlers of New Amstel—First elders and deacons—Dutch soldiers desert to Maryland—Council of New Amstel request Marylanders to return them—Baltimore claims South river—Utie sent by Maryland to demand it—Letter from Josiah Fendall, the Governor—Continued flight to Maryland and Virginia of settlers and soldiers—Stuyvessant disapproves of Alrick's course—Arrival of Utie at New Amstel—Demands the South river—Answer of Alricks and Beekman—Utie's threats—Firm and conciliatory answer of the Dutch—Information sent to Stuyvessant—He blames Alricks and Beekman for not arresting Utie as a spy—Appoints Martin Krygier and Van Ruyven to regulate affairs on South river—Krygier appointed Captain of the troops—Sixty soldiers sent from Manhattan—Commanded to arrest Utie as a spy—Augustus Herman and Resolved Waldron Ambassadors to Maryland—Their instructions—Arrival of Van Ruyven and Krygier at South river—They censure Alricks—Report of the deplorable condition of New Amstel—Tyranny of Alricks—Citizens refuse to enlist under him—Manhattan dissatisfied at sending soldiers to defend New Amstel—Complain of the number at the Whorekills—Directors in Holland disapprove of Alrick's conduct—Think it will ruin the colony—Again recommend disarming the Swedes, and compelling them to reside among the Dutch.

[1659] THE unfortunate town of New Amstel still suffered from the ravages of the fall fever, and to add to the misfortunes of Alricks, his wife died from the disease. In a letter describing the distressing condition of affairs, he said, "Winter early and long, and unexpected, caused great distress; the previously long continued rains prevented the collection of fodder for the creatures, and continued sickness curbed us all so far down, that all the labor in the field and agriculture was abandoned; the guns are rusty, not having any proper place to keep them in. One reason for the want of victuals is that the lands are new. I did see from the first, that from the New Netherlands settlers, who actually resided here at our arrival, scarce one obtained during our residence one schepel of grain; those who came with us hither, or emigrated afterwards to this place, did not much more, nor could effect anything better, as the time in the first year was spent in building houses and making gardens, in which small compass of garden each individual, as well in clearing soil, in building and carrying the materials, was so busily engaged that the summer was passed without having sown much seed in the ground; beside this, was then obstructed by the general prevailing sickness during two successive years, while the immoderate hot weather was another impediment."[1]

The desire of the Swedes to remain neutral in case of a war between Holland and Sweden, appears to have excited the distrust of the West India Com-

[1] Albany Records, vol. 12, pp. 480, 483.

pany. Accordingly, on the 13th of February, they wrote to Stuyvessant approving of all his orders except the appointment of Swedish officers. They said "the Swedes were not to be trusted." They told him "that it would have been preferable to have disarmed the whole nation, than to provide them with officers, and place arms in their hands which they might use against them, not only by the arrival of any Swedish succor, but on any other occasion." They told him "not only to remove the Swedish officers, and replace them with Dutch officers, but on the first favorable opportunity, to disarm them at the least symptom of disaffection." He was also instructed to endeavor to separate them, and induce them to settle amongst the Dutch inhabitants; and to admonish Alrick from time to time of his duty, and particularly to assist Beekman, who was continued custom-house officer and auditor of the colony of the city on the South river.[1]

At this time there were several alterations made in the conditions upon which the colonists had agreed to emigrate, by the Burgomasters of the city of Amsterdam.

The principles were as follows:

Provisions were only to be distributed from the public magazines, amongst those who had left Holland prior to December 1658. Merchandize was to be sold only for cash, and the city of Amsterdam was no longer obliged to keep supplies in their magazines. Exemptions from tenths, instead of continuing for

[1] Albany Records, vol. 4, pp. 291, 292.

twenty years, were to cease in 1678; and poundage, horn and salt money, ten years earlier than stipulated, "when taxes were to be imposed by the director according as the enclosed lands are situated near or at a distance." Goods in future were to be consigned exclusively to the city of Amsterdam, whereas the West India Company allowed all traders on South river to export whatever they pleased, except beavers and peltry, the monopoly of which was still retained by the city.

The promulgations of these new regulations caused intense consternation and dissatisfaction amongst the citizens of New Amstel, and this was not a little increased by the alleged tyranny of Alricks. A writer describing the effect it had upon the citizens of New Amstel at the time, says:

"Many poor folks, whilst they had anything left wherewith to pay for their passage, had offered it to Alricks, and besought him with clasped hands to accept it in payment for their debts, but he declined, saying, 'Ye are bound to remain four years.' 'We have spent in our hunger and wretchedness and misery all that we have saved from our small pittance. We have nothing left wherewith to pay,' was their reply. 'You must pay first, and then go,' was the answer of Alricks."

Numbers fled to Virginia and Maryland, where they spread the news of the weak and desperate condition of New Amstel.[1]

Stuyvessant, in a letter dated 4th of September,

[1] O'Call. vol. 2, p. 376, 377.

complains of this conduct of Alricks to the company.

The following fragment of a letter from Alricks to Stuyvessant, show some of the causes operating against the colony of New Amstel. He says:

"That prevailing violent sickness, which wasted a vast deal of goods and blood from one year to another, and which not only raged here, but everywhere throughout this province, and which consequently retarded not only our progress in agriculture, but threw a damp over the other undertakings. Besides that, in the ship 'Mill,' which only lately arrived, a very short time before the severe cold weather, were embarked more than two hundred souls, besides those who last spring arrived, and bringing, as appeared by the lists, about five hundred souls, without bringing any victuals with them, which baffles in this respect all our measures. It is true that we received by said ship a small cargo, about 3000 guilders worth for the purchase of victuals. The ship Mill arriving late, the harvest, by the unfavorable season being collected late, the little grain that was not drowned by the heavy incessant rains, but remained stifled in its growth, was sold at such excessive prices that it often could not be purchased where it was necessarily wanted. We were not permitted to go to Virginia, nor to the North, so that our bread magazine, our pantry room, our only refuge is to Manhattan."

Alricks despatched a galliott to Manhattan for food, but it was frozen up. A supply, however, was sent by Stuyvessant in the yacht Brigantine, consisting of pork, beef and maize, but she was run away with by her captain, Lumis Obbes, and the supply never reached the suffering colonists at New Amstel. Obbes went privateering. In the meantime the sickness still raged there. Alricks in his letter says, "sickness and death pressed upon us with such unabated violence, that a large number of men, and not a small number of our cattle perished."[1]

Agreeable to instruction, Beekman purchased from the savages the land from Bombay Hook to Cape Henlopen, named the Hoernkill. He departed for that purpose in company with D'Hinoyossa, on the 24th of May, and by the 14th of June had succeeded in completeing the purchase from the Indians. This was the third time that the Indians had sold the most of this land. They had first disposed of it to Godyn and Blommaert, and it was under the title from this sale that the settlement was made by the unfortunate first settlers of Delaware, who were massacred at Lewistown. They then sold it to the Swedes, and now they again sold it to Beekman and D'Hinnoyossa.

About this time one of the Swedish ministers attempted to preach in the City's Colony—in the town of New Amstel. The commissioners of the colony would not permit this on account of the difference between the religious faiths of the Dutch and Swedes. In a letter to Alricks they say: "The bold undertaking of the Swedish parson to preach in the colony without permission does not greatly please us. No

[1] Albany Records, vol. 12, pp. 484, 485.

other religion but the reformed, can or may be tolerated there, so you must by proper means put an end to, or prevent such presumption on the part of other sectaries."[1]

In a letter dated August 16th, to the Commissioners at Amsterdam, Alricks gives the following unflattering account of the settlers at New Amstel:

"In the Prince Maurice," said he, "were 35 colonists, free handicraft's men, amongst them some workmen, but the major part tradesmen, who did not learn their trades very well, and ran away from their masters too early, in consequence of their own viciousness. Also 47 soldiers, 10 civil servants, 76 women, children and maid servants. Those who arrived in the vessels De Waig, De Sonne, De Meulen, were of no good repute, scarcely three good farmers among the whole lot. The total was 137 tradesmen and servants, 70 soldiers and civil servants, 300 women and children, and the maid servants of the married women and children, &c., who came here as single women."

Alricks objected to this description of colonists, and desired "stout growing farm servants," and that the "women and children be omitted for the present, as agriculture could not be advanced without good farmers and strong laboring men."[2]

Two elders and two deacons were elected at this time in New Amstel. These were the first elders or deacons we have any account of in this State.[3]

[1] Broadhead and O'Calligan, vol. 2, p. 61.

[2] N. Y. Doc.; Broadhead and O'Calligan, vol. 2, pp. 68–71.

[3] Ibid.

The Dutch were now alarmed by the encroachments of the English from a new quarter. Heretofore their trouble was from the English of New Haven. It was now to commence with the English from Maryland. Amongst the many Dutch that had fled from New Amstel to Maryland, were six soldiers, who had deserted from the Dutch service. The Council of New Amstel held a meeting on the 20th of June at which it was resolved to request Josiah Fendall, the Governor of Maryland, to send these soldiers back. Being ignorant of the governor's address, on the 25th of June they sent the letter to Colonel Nathaniel Utie, (called by the Dutch Jude Utie), who was the chief of the Maryland magistrates, who resided, according to Dutch accounts, on Bearson Island, and solicited him to forward the letter. This Utie agreed to do, but at the same time informed the messenger that he had a "commission in his house to go to New Amstel," but that "in the meantime Lord Baltimore had arrived, and had commanded that the lands between the degrees of his grant should be reviewed and surveyed, and when ascertained, be reduced under his jurisdiction, without the intention of abandoning any part of it."

This being reported at New Amstel, together with the rumor that intruders on Baltimore's land were to be warned off, caused great anxiety amongst the inhabitants. Business operations were discontinued, and many prepared for flight.[1]

In accordance with this determination of Balti-

[1] Albany Records, vol. 13, p. 498.

more, a meeting of the Council of Maryland was held on the 3d of August, (old style), at Anne Arundel. Those present were Josiah Fendall, the governor, Philip Calvert, brother to Lord Baltimore, the Secretary, Col. Utie and Mr. Edward Lloyd. According to the minutes of the Council, "Then was taken into consideration his Lordship's instruction and command to send to the Dutch, in Delaware Bay, seated within his Lordship's province, to command them to be gone, and ordered that Colonel Nathaniel Utie do make his repair to the pretended governor of a people seated in Delaware Bay, within his Lordship's province, and that he do give them to understand that they are seated within his Lordship's province, without notice given to his Lordship's lieutenant here, and require them to depart the province."

"That in case he find an opportunity, he insinuate into the people there seated, that in case they make their application to his Lordship's governor here, they shall find good conditions, according to the conditions of plantations granted to all comers into this province, which shall be made good to them, and that they shall have protection in their lives, liberty and estates which they shall bring with them.

(Signed) PHILIP CALVERT."

The following letter was addressed to the "Commander of the people on Delaware Bay." From the tenor of it, it would appear to be in reply to one written to the governor by Alricks; the date of neither is given. But by the records of Maryland it is inserted under the proceedings of August 3d—it was, more than probable, written on that date. It says:

"SIR:—I received a letter from you, directed to me as the Lord Baltimore's Governor and Lieutenant of the Province of Maryland, wherein you suppose yourself to be the governor of a people seated in a part of Delaware Bay, which I am very well informed lieth to the southward of the degree of forty, and therefore can by no means own or acknowledge any for governor there but myself, who am by his Lordship appointed lieutenant of the whole province lying between these degrees, 38 and 40, but do by these require and command you to presently to depart north of his Lordship's province, or otherwise desire you to hold me excused if I use my utmost endeavor to reduce that part of his Lordship's province unto its due obedience under him."[1]

In the meantime, the affairs of New Amstel were so badly managed by Alricks, and his strictness, or rather tyranny, so great, that numbers of the inhabitants deserted the colony and fled to Maryland and Virginia.

The captain of an English ketch that had sailed from Boston with provisions, informed Stuyvessant that fifty persons, amongst whom were several families, had removed from New Amstel to Maryland and Virginia within a fortnight. Alricks even endeavored to get Stuyvessant to return those who fled from New Amstel to Manhattan; he was not even willing to accept pay and security for what they owed

[1] Maryland Records, Council, &c., H. H., 1656 to 1668, p. 43.

the city of Amsterdam, but insisted on their return. Stuyvessant in his letters to Holland severely censured this conduct of Alricks, and refused to return the fugitives.[1]

In the meantime, the desertions from the unfortunate colony of New Amstel to Maryland and Virginia still continued, until scarce thirty families remained. Of the fifty soldiers originally sent there, nearly one-half deserted, and only about eight or ten of them were garrisoned at New Amstel. The rest were sent to the Hoernkill. Thus, as Stuyvessant said, "leaving them in fear and peril of being massasacred by the cruel savages."[2]

Whilst the City's Colony was in this trouble, both soldiers and citizens deserting, Baltimore sent messengers demanding that the Dutch should abandon their settlement and jurisdiction on the South river, as they were within the limits of his grant, between the 38th and 40th degree of north latitude. The embassy was composed of six persons, viz.: Colonel Nathaniel Utie, his brother, his cousin, Major Jacob De Vrientz, and a servant. They brought with them four fugitives, of whom three were apprehended and one escaped. They arrived at New Amstel on Saturday, the 6th of September, and demanded an audience on the following Wednesday, which was consented to. At this meeting both Alricks and Beekman were present.

Utie first delivered his letter to Alricks, and then a

[1] Albany Records, vol. 18, pp. 28–29. [2] Ibid, p. 445.

copy of his instructions and his orders from the Governor of Maryland. He told them that the South river was in Baltimore's jurisdiction, and commanded the Dutch "to leave it directly, or declare themselves subject to Lord Baltimore." He also told them that if they "hesitated to resolve upon it voluntarily he deemed himself not responsible for the innocent blood that might be shed on that account."

The Dutch answered that his "communication appeared very strange in every respect, as they had been in possession of the land so many years, as well by an octoroy of the State General and the Directors of the West India Company, which they had previously obtained."

Utie replied that "he knew nothing about that." That "the land was granted to Lord Baltimore, and confirmed by the King himself, and renewed two years ago, and sanctioned by the Parliament to the extent of forty degrees." Utie then repeated again, "that he was innocent of the blood that might be shed, as Lord Baltimore was invested with the power of making war or peace without any man's control." He also said, "we ought to take hold of this opportunity, as the men had chiefly deserted New Amstel, and those who yet remained would be of little or no aid." He declared it was their "intention to take hold of this occasion, and not to let it pass," convinced as they were of the weakness of the Dutch. That "the present time suited them the best of the whole year, as the tobacco was chiefly harvested." Utie therefore demanded a positive answer, intimating at the same time it was indifferent to him how "they might resolve."

Alricks and Beekman answered that they "could not decide the case, but that it must be left to the lords spiritual and temporal in Holland and England."

Utie replied that he did "not care anything about them."

Alricks and Beekman then answered that they "could do nothing without them, and were in duty bound to refer the case to Stuyvessant, to whose government they were subject, and that it would require some time to consult him."

Utie asked "what time would be required?"

The Dutch proposed three weeks; on which Utie said, "I have no orders to give any respite, nevertheless, I will give you the required time."

On the 9th, Utie was again summoned to the fort to receive the Dutch answer in writing, when seeing Beekman, he addressed him particularly, telling him he understood that he "was commander at Christina, that he too must depart from there, as it was situated within the 40th degree of north latitude."

Beekman answered, "that if he had anything to say to him, he ought to appear at the place of his residence."

Utie replied, "I think it is sufficient, at all events, that I have made you this communication."

They then delivered to Utie a written protest, in which they state that the "instructions" (meaning the letter from the Council held at Annapolis on the 3d of August), given by Josiah Fendall, Lieutenant of Lord Baltimore, was without day, date or place, or where it should have been written, and was only signed by Philip Calvert, Secretary; that all related in relation to the alleged claim of Baltimore was unauthenticated by a single document. That his declaration that in case the Dutch refuse immediately to depart, he would be unaccountable for the innocent blood that might be spilled, appeared to them as "unexpected and strange treatment" by Christian and Protestant brethren, and near neighbors, with whom they desired and never solicited anything than a "sincere cultivation of harmony and friendship;" that they yet desire may be uninterrupted. They therefore requested at least an extract from the deeds and documents in relation to Baltimore's claim. In it they offered to show their title by grant from the State General, by transfer from the West India Company, and by payment made for the land and its actual possession. They desire that the differences might be settled by the States General and Parliament. They also protested against that part of Utie's instructions in relation to "the favorable terms and agreements about some plantatations to the inhabitants." They complained of the citizens of the South river being lured away to Maryland by promises of "protection and much liberty," some of whom were bound to their "lords and masters by oaths, and others who were in debt for considerable sums, by which their lords and masters are disappointed, and were frustrated to recover their debts." They protested also against the losses and damages they had already suffered, and might thereafter sustain, with a view of recovering compensation for such injuries thereafter. They also pointed to the treaty of alliance concluded on the 5th of April, 1654, between England and Holland, as well in America as Europe, "whereby they were charged and recommended to commit no hurt, hostility or injury against one another, as expressed by the 16th article."

This protest[1] was signed by the Director Generals or Governors Alricks and Beekman, and by the Council and Schepens, viz.: Alexander D'Hinoyossa, John Williemsen, John Crato and Hendrick Ripp, and by Secretary G. Vansweringen.

Immediate information was sent to Stuyvessant, overland, through the present State of New Jersey, of this visit and the demands of Utie, who in a letter dated the 23d of September, expressed his displeasure at what he termed "the frivolous fabricated instructions, without date or place" of Nathaniel Utie, and the "not less frivolous answers and proceedings with him, of the Governors and Council of Altona and New Amstel." He blamed them for allowing Utie "to sow," what he termed, "his seditious and mutinous seed among the community," during four or five days. Also for "for agreeing to give him an answer within three weeks, on his threatening expressions. This," Stuyvessant told them, "showed unquestionable proofs on their part of a want of prudence and

[1] See protest in full, in Albany Records, vol. 13; Hazard's Annals, p. 265; Holl. Doc. vol. 16, p. 117.

courage." He informed them they should "have apprehended Utie as a spy," and to show his want of confidence in them, appointed Captain Martin Krygier, a Burgomaster of New Amsterdam, and Cornelius Van Ruyven, his Secretary, to "dispose of and regulate the affairs" on South river, in relation to the proceedings of Baltimore and Utie. He also appointed Krygier commander of all the militia and soldiers on South river, and sent with him a reinforcement of sixty soldiers, to assist in protecting the Dutch settlements on the South river from invasion from Maryland. He also instructed them if Utie or any one else came back "for an answer for his frivolous demand, or frivolous signed promise, (such were the words of Stuyvessant), they were to arrest him as a *spy*, as not being entitled to an answer," unless "he exhibited a due qualification of a State Parliament or lawfully established government," and in the meantime hold him as a hostage until the Dutch might be acquainted (in the language of Stuyvessant), as to "where, how, and on whom" they might "take satisfaction for the cost and expenses they had already been at, or yet to be at, in the maintainance and defence of their own."[1]

At the same time two commissioners or ambassadors were appointed to visit the Governor of Maryland. They were Augustine Hermans and Resolved Waldron. Herman (or Harman, as the name was afterwards changed to), was the first proprietor of the

[1] A. P. MSS. in Reg. of Penn. vol. 4, p. 98; Hazard's Annals, pp. 267, 268.

celebrated Bohemia Manor, consisting of eighteen thousand acres of land, which lays partly in St. Georges and Pencader hundreds, in New Castle county, and partly in Cecil county, Maryland. This land is supposed to be the best in Delaware. Herman[1] (from whom is descended several of the most celebrated families of this State, some of whom still possess the land, derived by descent from him), was originally from Bohemia, and when he came to New Amsterdam was clerk to John and Charles Gabry, of Amsterdam, in Holland. He was a man of great ability, and amongst his other qualifications, was a good surveyor and draughtsman.

In 1647 he was appointed by the Director and Council of New Netherlands, one of the nine men, a body of citizens selected to assist the government by their counsel and advice. His first wife was Janneken Verlett, of Utrecht, whom he married in New Amsterdam, December 19, 1650. He was formerly opposed to Stuyvessant in the disputes that divided New Netherlands.

Adrian Van Tienhoven, on the contrary, formerly Secretary of the Colony on the South river, was a firm supporter of Stuyvessant, whilst Herman and Van Dincklage, Govert Lockerman Van Derdonk, and others, formed a combination against him. Herman and Van Dincklage wrote several letters to Holland, severely condemning his conduct. The following is an extract from one of Herman's

[1] A more full account of him will be found in the following portions of the history.

to Van Derdonk,[1] yet extant, dated September 20, 1651:

"Govert Lockerman is totally ruined, because he will not sign that 'he knows and can say nothing of Director Stuyvessant but what is honest and honorable.' I fear we too shall experience a like fate, whether we have safeguard from their High Mightinesses or not. 'Tis all alike. The Directors have written not to pay any attention to their High Mightinesses' safeguard or letters, but to theirs; and every one can see how prejudicial that is to us. We are turned out, and dare scarcely speak a word. In fine, matters are so situated that God's help only will avail. There is no trust to be placed in man. That infernal swaggerer (blassegust) Tienhoven, has returned here and put the country in a blaze. Things prosper, they report, according to their wishes, to which I know not what to answer, &c. The basketmaker's daughter, of Amsterdam, whom he seduced in Holland, on a promise of marriage, coming and finding that he was already married, hath exposed his conduct even in the public court. Your private estate is going all to ruin, for our enemies know how to fix all this, and attain their object. There is no use in complaining. We must suffer injustice for justice. At present, that is our wages and thanks for our devotion to the public interests. Yet we will trust in God."[2]

[1] Van Derdonk was banished from the Colony for seven years, for abusing the Directors.

[2] The basketmakers' daughter was a girl named Lisbeth Van Hoogvelt, whom Tienhoven debauched and lived with as his wife, whilst in the Hague, having at the same time a wife in New Netherlands. This affair cast a stain on the character of Tienhoven.

Van Dincklage also writing to Van Derdonk in a letter dated the 19th of September, 1651, speaking of Stuyvessant, says:

"Our great muscovy duck goes on as usual with something of the wolf. The older he gets, the more inclined he is to bite. He proceeds no longer by words and writings, but by arrest and stripes."[1]

The dissensions of which the above letters are exemplifications, at this time must have been healed, or Herman would never been appointed to the responsible station of ambassador by Stuyvessant.

Herman and Waldron took with them a letter from Stuyvessant to the Governor and Council of Maryland, in which he expressed his astonishment at the arrival of Col. Utie at New Amstel, and the demand for the South river, and complained greviously of his conduct "in threatening the government, council and inhabitants of that place with blood," in case the territory was not given up within three weeks; and also of his "having sought to alienate and induce to rebellion from their lawful commander, the citizens of New Amstel." He at the same time instructed the ambassadors in a "friendly and neighborly way" to request the Governor of Maryland to deliver up to the Dutch "such free people and servants as for debt and other ways" had fled, and taken refuge in Maryland. If this was done, they were instructed to agree that all runaways and fugitives from Maryland, or South river, should also be delivered up to the Maryland authorities, and that they would in every way

[1] Broadhead and O'Calligan, vol. 1, p. 453.

"maintain good justice and neighborly duty." If the Governor and Council of Maryland refused this, the ambassadors were to inform them that the law of retaliation would be enforced, and that they in return would refuse to deliver up "all servants and negroes that might escape from Maryland to the South river."

In relation to the demand made for the surrender of South river, through Col. Utie, they were to represent that "threats to invade by way of hostility, any possession of the Dutch on South river, was altogether contrary to the 2d, 3d and 16th articles of the confederacy of peace made between the Republic of England and the Netherlands in 1654."[1] That the Dutch had had possession of the South river (by grant from the Lords State General, and by deeds of the natives), for over forty years. That by these articles of the treaty between England and Holland all questions in dispute were to be referred to their decision. The ambassadors were therefore by virtue of those articles of peace to demand of the Governor and Council of Maryland to give them reparation and damages against Nathaniel Utie for his "frivolous demands and bloody threatenings."

Van Ryven and Krygier, who had embarked from Manhattan, with their troops in three barks, on the 23d of September, arrived at the South river, opposite New Amstel on the 28th. They found the people in commotion and fear respecting the threatened

[1] The English government from 1649 to 1661 was Republican under the rule of Oliver Cromwell, styled the Protector of the Commonwealth of England.

English invasion. Upon examining into affairs, they saw much to condemn in the conduct of Alricks. They charged him as the cause "of all the misfortunes in New Amstel." "In such bad name is this place," (New Amstel), said they, "that the whole river cannot wash it out off, and would to God that it remained here, and that it was not openly proclaimed in our fatherland, and to the scorn of this whole province." They denounced him for oppressing the people. For first refusing them liberty to leave New Amstel when they offered to pay the debts they owed the city, but insisting on their remaining four years, and afterward when their money was spent, and they were sick and hungry, not allowing them to leave until their debt to the city was satisfied. It was reported they said, "that many actually died by hunger." So unpopular had Alricks become that the citizens would not enlist under him for the defence of New Amstel, although they were willing to engage under Krygier. And when Utie was there making his demand for its surrender, and Beekman proposed to detain him, Alricks declined to do so, for fear of a revolt of the citizens. Van Ryven in writing to Alricks afterwards, charged him with making no effectual means to enlist men. "Did one of the city officers stir one single step towards this object?" "Or shall it be urged that it was published by beat of drum: But no person enlisted?" wrote Van Ryven. "It was known beforehand that none could be obtained in this manner," he asserted, "at least not from the inhabitants, because the great majority who yet remain in the city's service are dissatisfied with the magistrates of this colony, for what reason," says Van Ryven," "must be best known to your honor." "These persons," said he, "ought to have been encouraged by favorable terms and salary, as is the usage in fatherland, and anywhere else, in such great distress." At this time the Dutch, in addition to their troubles with the Marylanders, expected a war with the Indians at Manhattan, and endeavored to get Alricks to enlist fifty men. They were dissatisfied at taking the soldiers from Manhattan to defend New Amstel, which they thought should be able to defend itself. They also condemned the sending of so many of the soldiers to Hoernkill.

In speaking of this matter, Van Ryven in a letter to Alricks says:

"But what excuse can be made why the soldiers on the Hoernkills, as we were promised last September, were not commanded to march hither or have not arrived. It is indeed too absurd, that the Director General and Council should believe their own places of far greater consequence of the necessary soldiers, and send them hither for succor, and that you should not send for your own soldiers, but leave them to guard one or two houses, built apparently more for private views than for the welfare of the country, and employ sixteen or eighteen for this purpose."[1]

The Directors of the West India Company when they heard of the conduct of Alricks in relation to the oppression at New Amstel, expressed their disapproval in a letter to Stuyvessant. They considered

[1] See Van Ryven's letter, vol. 18, p. 425, 426, N. Y. Records.

"it a symptom which threatened a total ruin of the colony," and that gave no prospect of a return for the expenses that had been entered into. They laid the whole to the "too rigid preciseness" of Alricks, in not permitting the New Amstel colonists to settle at Manhattan. Stuyvessant was instructed to try to divert Alricks from "this plan as soon as possible." He was told to show to Alricks that "at this critical moment it would be far preferable, if he would make voluntarily an offer to the remaining creditors to settle in the Manhattans, provided they gave bonds for the debts which they were yet owing." In "this case," say they, "their recovery may sooner or later be expected, which is utterly hopeless if they remove from the district of the company, and settle anywhere else." He was also instructed not to compel any New Amstel colonist to return who had settled at Manhattan. Also to solicit the return of those who had settled in Virginia and other neighboring districts, and employ every feasible means to that end. They also informed him that they persisted in the sentiment that the Swedes should be separated one from another, and if possible amalgamated with the Dutch nation, and that they should be disarmed at the earliest opportunity. Stuyvessant was recommended to do this before they (the Swedes) could make any alliance with the English to the disadvantage of the Dutch. The possibility of this alliance appears to have given considerable uneasiness both to Stuyvessant and the Directors in Holland.

The Indians about this time killed four of the Dutch settlers. On this account Alricks and Beek man had great difficulty in sending information of Baltimore's threatened invasion to Manhattan, no one being willing to cross through New Jersey from New Amstel to Manhattan by land.

Alricks in a letter to De Graeff, Burgomaster of Amsterdam, who appears to have had the main charge of the affairs of the City's Colony says: "that there were one hundred and ten houses in New Amstel, and 16 or 17 more on land belonging to our nation, and 13 or 14 belonging to the Swedes." He does not mention whether these other houses are situated in the City's Colony, or the Company's Colony of Altona, north of the Christiana. But we infer from the tenor of his letter that he means that this number of houses were in New Amstel and its neighborhood alone. As he intended to give to the agent of the city of Amsterdam a description of their property, and not of that of the Company. He is incessant (in his letter to the New Amsterdam Commissioners) in his demand for practical farmers. He especially desired 20 or 25 families of good agriculturists, and 30 or 40 cows to each family. The cattle he recommended should be furnished by the city of Amsterdam, to be kept on shares, the family to have half, and the city half. He describes the goods as most in demand at New Amstel, as duffels, gray osnaburgs, and strong liquors. "Clearing lands," he says, "furnishes considerable employment here. Plowing, sowing, mowing and thrashing, requiring strong people, accustomed to labor, most of whom should, as far as possible be men."[1]

[1] Broadhead and O'Calligan, N. Y. Doc. p. 76, 78, vol. 2.

CHAPTER XXII.

FROM 1659 TO 1660.

Herman and Waldron leave as Ambassadors to Maryland—Difficulties in their way—Threatened by two Fins—Stop at Capt. Wick's plantation—Are informed that the English believe the Dutch incited the Indians to murder them—Visit Secretary Calvert—Conversation with Calvert on English and Dutch rights—Meet Governor Fendall and Maryland Council—Claim for the Dutch the territory between the 38th and 42d degree of north latitude—Recite the Dutch claim—Complain of Col. Utie—Demand return of fugitives—Deny claim of Maryland—Marylanders justify Col. Utie—Persist in their claim—Endorse his instructions—Marylanders claim between 38 and 40 degrees latitude—Extent of said claim—Warmth of Col. Utie—Discussion between him and the Ambassadors—Marylanders demand to see the Dutch patent—They have not got it—The mountain—The Ambassadors depart—Herman goes to Virginia—Dutch suspicious of plots—Beekman sick—Stuyvessant fears the English—Urges population—Disputes between the authorities of City and Company—Recrimination between them—Death of Alricks—D'Hinnoyossa appointed his successor—Proposition to tax the Swedes.

On the 30th of September, Herman and Waldron, the ambassadors appointed by Stuyvessant, left New Amstel for Maryland. They were accompanied by some guides, mostly Indians, and convoyed by a few soldiers. They travelled by land, taking the first day a course W. N. W. from New Amstel. They continued this course for 4½ Dutch miles, (about 13¼ English), when they took a due west course, and after travelling three more Dutch miles, the Indians refusing to proceed any further, encamped for the night. On the 1st of October they continued their travel, going W. by S., and then again directly South. The country at first was hilly, and then low. They soon arrived at a stream, which the Indians informed them flowed into the Bay of Virginia, the Chesapeake Bay. They followed this stream until they found a boat hauled upon the shore, and almost dried up. Dismissing four of their guides, and retaining only a man named Sandy Boyer, and his Indian, they pushed off, but were soon obliged to land again, as the boat became full of water, whereupon they turned the boat upside down, and caulked the seams with old linen. They thus made it a little tighter, but one was obliged to sit continually and bail out the water. Proceeding down this stream, they soon reached the Elk river. Here they made a fire, and proceeded with the evening tide, but with great trouble, as the boat had neither rudder nor oars, but only paddles. Having rowed all night, on the 2d of October they reached the Sassafras river, and stopped at the plantation of a man named John Turner. Here they met a man named Abraham, (a Fin), a soldier of Altona, who had run away with a Dutch woman. They were offered a pardon if they would return to Manhattan or New Amstel within a month. The woman agreed to do this. She had three months of her time to serve—but the soldier refused; he, however, made two oars for them. Here they set Sandy Boyer on shore for information, but they could get none, as the only inhabitants were a few Fins and Swedes who had deserted from the South river, in the time of Governor

Printz. They proceeded onward, but had scarcely left the shore when this Abraham and another Fin named Marcus, followed them in a canoe, and endeavored to stop their passage, claiming the boat as their own. Herman and Waldron desired to proceed, assuring them they should have the boat on their return; but they endeavored to stop them by force, and Marcus drew a pocket pistol, and threatened to fire. They also had two guns with them. At last, with difficulty, they succeeded in getting rid of them. At the mouth of the Sassafras they came to Colonel Utie's, where they heard a "strong firing," supposed to proceed from fifty to sixty men. They supposed from this that an expedition was being prepared to visit South river. On the 3d of October they entered the Chesapeake Bay, and in the evening arrived at the plantation of Captain Wicks, one of the three magistrates of Kent Island. Of him they endeavored to learn whether the English had laid any regular plan for attacking South river. Wicks informed them he understood "it belonged to Lord Baltimore, and that he was obliged to sustain him in his right and title." Herman and Waldron endeavored to prove the contrary, and informed Wicks that "he who would have it, must get it by force. That the Dutch had already sent one hundred men from New Amsterdam for the defence of South river, and that double that number were soon expected," but they hoped to be on friendly terms.

The Dutch here learned during the discussion with Captain Wicks that the English were informed that the Dutch at Hoernkill were stirring up the Indians in the war they were then engaged in against the English. That it took place thus:

"A certain savage met a Dutchman at Whorekill,[1] and told him he would kill a Dutchman, because his father had been killed by a Dutchman before. To which the Dutchman replied that his father had been killed by an Englishman, and therefore ought to take revenge on them. On which the savage went off and killed an Englishman. Thus the war, said one of the English, was aided by the Dutch in supplying arms."

Herman at first denied this, and then palliated it.

Of Captain Wicks the embassy procured a boat, and sailing down the Chesapeake, on the 7th of October they arrived at Secretary Calvert's. They were quartered in the neighborhood, on a Mr. Simon Overfee or Overzee.

On the 8th of October they invited Calvert, the Secretary, to dine with them, when they had a familiar conversation on the affairs of South river and Maryland.

Calvert during the conversation said he "wished happiness to Maryland and Manhattan."

"This," the ambassadors remarked, "included the whole land, it having retained its ancient name from the tribe of savages among whom the Dutch made a beginning of the first settlement."

[1] This is spelt by the Dutch Hoernkill and Hoerkill. When giving Dutch accounts, we shall spell it in the former way; when English the latter.

Gradually they struck on the point of the limits, which Calvert said of Maryland, was "between 38 and 40 degrees of latitude along the sea, by which Delaware Bay was included, and then in a direct course to Paman's Island, and thence to the origin of Potomac river."

The ambassadors answered the 38th and 40th degrees ought to be understood of the Chesapeake Bay upward, and then the colony of Virginia reached, the same bay to the sea.

Calvert replied "that it was not so; but that it ought to meet the limits of New England."[1]

"Where, then," exclaimed Herman and Waldron, "would remain New Netherlands, if their limits were to join New England?"

Calvert answered, "he did not know."

The ambassadors then said that they "knew for both of them together, that it was a mistake, and that New Netherlands was in possession of those limits several years before Lord Baltimore obtained his patent, and they actually settled those spots." The ambassadors further alleged that "Edmund Ployden had in former days made a claim to Delaware Bay, and that the one pretension was not better supported than the other."

To this Calvert replied "that Ployden had not obtained a commission, and was in England thrown in

[1] The charter for New England, granted in 1606, called originally for all the land between the 41st and 45th degrees of latitude. In 1620 the 40th degree was added, thus effectually granting away all the lands settled on by the Dutch. See ante pp. 89, 91.

jail for his debts. That Ployden had solicited from the king a patent for New Albion, but had been refused, on which he addressed the Viceroy of Ireland, of whom he obtained a patent, but that was of no value at all."

The ambassadors, to use their own words, then said, "On this we entirely confounded him with his own words, by saying that it could not be known of my Lord Baltimore's pretensions, if he had any on the Delaware Bay, had obtained these by false or foreign representations. Neither could it be believed that the King of England, who once took notice of the Dutch plantations in New Netherlands, and who commanded those of Virginia and New England, as we could prove by their own English authors, expressly to remain at a distance of one hundred leagues from one another, determining nothing about it. It was therefore an unquestionable proof that he might reach the borders of New England, that it then was void, and of no value whatever."

The ambassadors were not able to see the governor for several days, and accordingly on the 12th of October, Calvert was again invited to dine with them, when, after the cloth was removed, another discussion took place as to the relative claims of the Marylanders and Dutch to the South river. Three maps were introduced. One printed at Amsterdam by direction of Captain Smith, the first discoverer of the Chesapeake Bay; the second also printed at Amsterdam, about the time of Baltimore's patent. The other was in manuscript. They all differed from each other.

By these maps, said Herman, "they endeavored to prove Lord Baltimore's claim, but we showed that the Bay of Chesapeake being so much to the northeast would come in our limits."

"How can that be, as New England was discovered first?" exclaimed Calvert.

"On this," says Herman, the Dutch alleged "they were nearly three years earlier in their parts than the English were in theirs."

Calvert replied, "they counted from Walter Raleigh."

"We," said Herman, "derived our right from Spain."

Calvert in reply said, "the Dutch were then not a free nation."

"Waxing warm," said Herman, "we took up other subjects."

The ambassadors stopped at Mr. Overfees until the 16th of October, when they were informed that Governor Fendall would meet them at the house of Mr. Bateman, at Potusk. Two horses were sent for them. They arrived at Mr. Bateman's about 3 or 4 o'clock in the afternoon. They were courteously received by the governor, and invited to dinner. After dinner the governor gave them an audience. Herman was placed on the left, Secretary Calvert on the right; then followed Waldron and the members of the Council, who consisted of Captain William Stone, Thomas Gerrard, Luke Barber, Colonel Nathaniel Utie, Baker Brooke and Edward Lloyd.

The ambassadors then made, or rather read, a communication, copies of which, in Dutch, were handed to the English, and afterwards rendered into English by Mr. Overfee.

They first cited their claim as by grant from the States General of the land between the lines of latitude of 38 and 42 degrees. The extent of this claim they do not appear to have been aware of. But it included the territory at present comprised in the Eastern Shore of Maryland, the State of Delaware, New Jersey, Connecticut and Rhode Island, and part of New York and Massachusetts, including the celebrated Plymouth Rock, the landing place of the pilgrims. They claimed this right first from the Spaniards who were the first discoverers of America, and of whom they were at the time subjects, and who afterwards they alleged, (when Holland separated from Spain, and became independent,) "granted to them amongst other territories, that at present called New Netherlands, now also secured by the right of possession and discovery." After reciting the various claims of the English and French, based on the right of discovery (making, however, no mention of Sebastian Cabot, who under the English flag was the first discoverer of the continent), they asserted that the boundaries of the possessions of the Christian Princes of Europe, were "by communication with each other's ambassadors agreed upon." That upon this agreement King James of England, "commanded and required that Virginia and New England should remain asunder and not meet together within the space of a hundred leagues," which space they alleged "was alloted for the Dutch plantations, then called by the general name of Manhattans, after the name of the Indians, they were first seated by." They scouted the idea that the name of Manhattan applied only to the island upon which New Amsterdam was built. The South river, they alleged was in the primitive times possessed by the Dutch, and "a colony planted on the western shore, within the mouth of the South Cape, called Hoernkill." That they "built there a little fort, and established a colony, but that it was massacred by the Indians." After this, in 1623, they "built the Fort Nassau, about fifteen leagues up the river, on the eastern shore, besides many other places built by the Dutch, and the Dutch Swedes." That they "thought it well to remove Fort Nassau downwards to the western shore, and there to fix a town," which they asserted "stood at this day, no man ever making any protest or claim from Maryland or Virginia against it." They also claimed "to have just right and title to the South river by lawful purchase from the Indians, the natural proprietors of the soil." They also alleged that from the "primitive time aforesaid, they had always held friendly and neighborly correspondence with the English of Maryland and Virginia, without any claim, injury or molestation to one another, until the 8th day of September, when Colonel Nathaniel Utie came into the town and fort of New Amstel, erected in the year 1650, and without any special commission or lawful authority from any state, prince, parliament or government, exhibited only by a piece of paper, or cartabel, by form of an instruction from Philip Calvert, Secretary, written without year or day, name or place, were neither signed or sealed by any state, prince, parliament or government, demanded in a manner, and required in a strange way, that the town and the country should be delivered up to the province of Maryland, as he said, for my Lord Baltimore; going from house to house to seduce and draw the inhabitants to rebel and fall from their right and lawful lords, and threatening, in case they did not submit and deliver up possession of New Amstel, to come again with force of arms, and fire and sword, and plunder them, and take their houses from them." Against this the ambassadors protested.

This conduct of Utie they asserted was "not only against the law of nations, neighborly friendship and common equity, but also contrary to articles 2, 3, 5, 6, 9, 10 and 16, of the treaty of 1654, between the two republics of England and the United Provinces. By the conduct of the said Nathaniel Utie, the said treaty of amity and peace was disturbed and interrupted, and they demanded justice and satisfaction for the wrongs and damages they had suffered and might suffer, according to the 16th article of the treaty."

They also demanded the sending back "of all Swedes and Dutch subjects, runaways and fugitives, who from time to time (especially the year 1659) had ran away from the South river. For the most part," they said "they were deeply indebted or delinquents." They declared that New Netherlands would also engage to return fugitives which might come into her jurisdiction. If not according to the *leges taglionis* (or law of retaliation), New Netherlands would hold itself "constrained, necessitated and excused to publish free liberty, access and recess to all planters, servants, negroes, fugitives and runaways, which from time to time might come into her jurisdiction."

The ambassadors utterly denied that Baltimore under his patent had any right to South river whatever. They asserted they had had possession of the territory for forty years, whilst the patent of Baltimore was of no longer standing and settlement than twenty-four or twenty-seven years. That he had not even as much title as Sir Edmund Ployden. That even if he had had a title to the South river, according to the 30th article of the treaty between England and Holland, he should have made his claim known before the 18th of May, 1652, to the commissioner appointed by both England and Holland to determine such differences as might have occurred between the year 1611 and the 18th day of May, 1652. In proof of the correctness of this, they stated that "New England having set up some claims to South river, and the ships of the Lords Protector (Cromwell) having been sent there to subdue the province of New Netherlands, upon peace being concluded between Holland and England, the design of the conquest of New Netherlands was given us, and the ships of war were sent against the French." They also said that their "western limits had been questioned, and having thereupon observed and suspected the Bay of Chesapeake, in the uttermost part therefore winding so much to the northeast as to run about Sassafras and Elk river, in our (their) line, they therefore laid claim to those parts, until by due examination hereafter the truth thereof may be found out, or agreed and settled amongst us (them) otherwise."

The ambassadors concluded by declaring that they never meant any "wrong or offence to the provinces of Virginia or Maryland," but on the contrary, still desired to continue in neighborly amity, confederacy and friendship; that they only demanded that "justice and satisfaction might be given." To prevent further mischief, they advised that three rational persons should be appointed from each province, to meet at a certain specified time, about the middle, between the Chesapeake Bay and South river, at a point which they described as "a hill[1] lying to the head of Sassafras river and another river coming from our river, almost meet together," with full power to settle the bounds and limits between the provinces of Maryland and New Netherlands. If such a settlement is not possible, then to refer the matter in dispute to their sovereigns at home, the States of England and Holland. But that in the meantime "all further hostility or infractions towards each other may cease," so that the soldiers that were sent to defend New Amstel might be sent home—and no further expense be added, and that a fair correspondence might be kept up between them as heretofore. After desiring that what they had stated might be recorded, they wound up their address by the following:

[1] This hill, we think, must be Iron Hill, near the village of Newark, in New Castle county.

"And so wishing the Lord God Almighty will conduct your honors both to all prudent results, that we may live neighborly together in this wilderness, to the advancement of God's glory and kingdom of heaven, amongst the heathens, and not to the destruction of each other's Christian blood, whereby to strengthen the barbarous Indians. Nay, may we rather join in love, and league together against them, which God and our Saviour will grant."

After this the commissioners[1] delivered a written copy and withdrew from the room.

The ambassadors made no allusion to Hudson, who discovered both the North and South rivers.

After their withdrawal, the Maryland Council took the matter into consideration, and resolved to have an answer ready by Saturday, the 8th, at 5 o'clock P. M. They then adjourned until the next day, when they again met, and "after a long debate, it being considered that Baltimore's instructions and orders were only to give the Dutch warning to be gone, and that when they were able to beat them out they might not plead ignorance, it was resolved that an answer in writing, by way of a letter, should be given, directed to the General of the Manhattans."

In accordance with this resolution a letter was prepared, in which they acknowledged letters of credence by the ambassadors, containing "many expressions of love and amity." It said they "felt themselves obliged to return them real thanks, in giving them (the Marylanders) an opportunity of unfolding

[1] Maryland Records, from the book entitled "Council, &c., H. H."

the causes that had been the reason of the Dutch astonishment and wonder." It said, it would give the Dutch that satisfaction which "with reason they could expect," and which they should likewise exact from them (the Dutch) as substitutes of Celius, Lord Baron of Baltimore, proprietary of that province, and that part lying on Delaware Bay, to them intrusted, and on which the Dutch had injuriously seated themselves, to the prejudice of Lord Baltimore's right and title. In answer to the ambassadors demand made on them, they said that Colonel Nathaniel Utie was by them "in pursuance of a command from Lord Baltimore, ordered to make his repair to a certain people seated on the Delaware Bay, within the 40th degree of north latitude, and let them know they were residing in his Lordship's jurisdiction without his knowledge, and much more, without his license, and without grant of land from, or oath of fidelity taken to his Lordship, which were the conditions and laws on which he granted the plantations to those within the jurisdiction of his grant. In case of their refusal he was instructed to let them know that all lawful means were to be used to reduce them to obedience, which all people within his territories were bound to yield to those entrusted with the province of Baltimore, who was sole and absolute lord and proprietary, by patents under the great seal of England, bearing date the 20th of June, 1632, and since confirmed by acts of Parliament," a copy of which was shown to the ambassadors. It said, "that as the ambassadors seemed to insinuate the colony on the Delaware was seated thereby and under the command of Stuyvessant, to whom the letter was directed as General of the Manhattans, they protested against him and all other persons, either principals or abettors, in the said intrusion upon their bounds and confines." It also declared their intention of recovering in due time, and by all lawful means, damages and costs that they might have, or hereafter might have sustained from the Dutch occupancy and injurious detention of the territory within their bounds and limits.

The letter also asserted "the original rights of the Kings of England to these countries and territories, which it was their (the Maryland Governor and Council's) business to maintain, and not by any discourse to controvert, or in the least attempt to yield up. That they could accept nothing from any other power, nor could they yield up any authority without the consent of the Kings of England and their successors. That the Dutch had no authority to exercise any jurisdiction on the Delaware by virtue of any grant from the State General, because the State General had no authority to make such grant, and if they did make one, it would be void and of no force and effect."

In relation to the instructions to Colonel Utie, (so much insisted on by the Dutch), they said "were such as every person inhabitant of this province of Maryland ought to take notice of." It was signed by the Secretary of the province, and this was the usual and common mode of giving notice to the inhabitants of Maryland, and that they made use of no other. It stated that they did not believe that the State General "would own those people at Delaware Bay to be there seated by their authority, since they have heretofore protested to the supreme authority then in England, not to own their intrusion upon their territories and dominions."

In relation to indebted persons, it informed the Dutch "that the Maryland Courts were open—that their justice was speedy, and denied to none that should demand it, which they thought was as much as in reason could be expected of them. That they took the same course in relation to the neighboring colony of Virginia, and only gave the Virginians and their brethren (the English) the same remedy."

It wound up by saying: "Thus hoping that you will seriously weigh the consequences of your actions, we rest in expectation of such a compliance as the style you give yourselves imports, having taught us to subscribe ourselves your affectionate friends and neighbors."[1]

Notice was then given to the ambassadors "to attend on the next day for an answer." Accordingly Herman and Waldron attended, and presented a paper, which stated "that having viewed Lord Baltimore's patent, in which they say they reserved only what the Governor and Council of New Netherlands might have to say against it; that they repeat their former declaration and manifestation of the 6th inst.,

[1] N. Y. Historical Col. vol. 3, pp. 384, 385.

and that Baltimore's patent, that he had petitioned the King of England for a country in the parts of America which was not seated and taken up before, only inhabited (as he saith) by a certain barbarous people, the Indians; upon which ground his royal majesty of England did grant and confirm the said patent. But that the South river, of old called the Nassau river of New Netherlands, (by the English surnamed Delaware), was taken up, appropriated, and purchased by virtue of a grant from the State General, long before the grant to Baltimore. Therefore the King of England's intention and justice was not to have given and granted that part of a country which before was taken in possession and seated by the subjects of Holland. So that the claim of Baltimore to Delaware Bay or any part thereof was invalid."

It was on these words of the grant Baltimore was was afterwards defeated by Penn, in his claim to this State—otherwise the State of Delaware, the City of Philadelphia, the town of West Chester, and portions of Delaware, Chester, York, Adams, Franklin, Fulton, Bedford and Somerset counties, Pennsylvania would now be included within the limits of the State of Maryland.[1]

In addition to these formal letters and written speeches, considerable conversation relative to the South river and the conduct of Utie took place. The governor asked the ambassadors "if his letter by

[1] Maryland Records, copied by J. Leeds Borzman for the New York Historical Society.

Utie had been received by the Director General and Council?"

They said "no, they had received no letter. That they were informed on South river that Alricks had received a private one, in answer to one of his without date, time, or place, of which he took no notice."

Governor Fendall said "he had no intention to meddle with the government at Manhattan, but that the government and people who had settled on the Delaware Bay, were within their limits, and that he once sent Colonel Utie to them, and that he should have delivered his instructions, though only given to regulate his conduct, &c."

The ambassadors replied "that the government and inhabitants on South river made separate government, but a sabaltern and subject, being only Vice Governors and Members of New Netherlands," &c.

Fendall answered "that he knew no better, and had always understood that the General Director on South river, in Delaware Bay, did hold his commission from the city of Amsterdam, and had settled there with his people as a separate government."

The ambassadors answered, "no; but that the city of Amsterdam was in possession of that place as a colony, and a particular member of New Netherlands, in a similar manner as their colonies in Virginia and Maryland were subsisting; and they had many similar colonies in New Netherlands, so that any injustice or injury committed against the colony of New Amstel, was perpetrated against the whole State of New Netherlands."

This answer of the ambassadors appeared to have offended Col. Utie, who said with great vehemence, "that they might take notice of all that had happened, but that all that he had done against the people who had dared to settle within the province of Lord Baltimore, if the Governor and Council would renew his commission, he would do again."

Herman and Waldron replied that "if he returned once more, and acted in the same manner as before, he would lose the name of ambassador, and be considered as a pertrubator of the public peace, because it was not lawful in an ambassador or delegate to attempt any other thing than to present in courteous manner his message to the magistrate or supreme chief to whom he was sent, but that it (his) was the language of open hostility, a language of war, to summons a place to surrender in such a manner as by fire and sword."

To this Utie answered that "he had not done so, further than his instructions and commission justified."

The ambassadors replied "that they would only pay regard to the answer which they received in return, and therein he would clearly perceive in what manner he made his."

Utie further added "that he too had further understood that they had threatened to transport him to Holland, which he wished they had executed."

They replied "that if he once more returned and acted in that manner, perhaps that nothing better might be his lot."

Utie then asked "in what manner he ought to have conducted himself; he had despatched two of his men before him," he said, "to notify his arrival, after which he took up his abode in the city, and if it then was not permitted to take a walk and look at the place, and converse with the inhabitants, who invited him to enter their lodgings."

The ambassadors answered "that it was permitted to do this, but not to stir up revolt and rebellion against the magistrates, and threaten them if they would not voluntarily surrender, that they were to be plundered and expelled, so that those altercations caused uneasiness on both sides."

Utie at this (the ambassadors said), glowed with rage, and was commanded by the governor to keep himself more reasonable. That they remained at full liberty to explain themselves without interrupting each other.

The ambassadors "then appealed to what they had brought with them in answer from New Netherlands, which they declared they had made known, and which they solicited might be taken into serious consideration, so that they might avoid any frivolous discourses."

Fendall hinted to Herman and Waldron, amongst other points, that they had "arrived there without having demanded or obtained, as they ought to have done, a license."

They remarked that they "were yet unacquainted with the forms of the government, but would conduct themselves in future in accordance to their custom, or such as they should deem proper to establish."

On this Utie exclaimed that they "should have stopped at his island to inquire there whether they should be permitted to land."

The ambassadors thought from this that if he had met them, or known anything about their embassy, he would have kept them there, and not permitted them to proceed further. But one of the council interrupting Utie, said, "that then they would have been accommodated with a better vessel," as Herman and Waldron had before told them they had arrived in an old and leaky boat, and that they could not wait to procure a better one. They however thought that if they had not exerted themselves to the utmost on the road to avoid Colonel Utie, he would have left nothing untried to disappoint them and frustrate their plan.

This was the end of this conference.

At another time they had friendly discussions with the Governor and Council individually. They "proposed to submit the matters to a committee of both nations, or enter into friendly correspondence for trade," &c. This, the ambassadors said they seemed to consent to, but they were inclined to defend their rights under the patent.

They also held an interview with Governor Fendall after they had given their answer in relation to Baltimore's patent, when he told them that Baltimore's patent was given by his majesty, with full instructions that Delaware Bay was to belong to the English. He also required of them the Dutch patent for New Netherlands and the Delaware Bay.

They answered that "they did not need to expose it at present, as they did not come for that purpose, but only to prepare a day for a future meeting between both parties."

Fendall then thought that he "ought not to have shown his."

The ambassadors then replied "they intended no other use for it than the Delaware."

Fendall said that "Claiborne had before made the same objection regarding the Island of Kent, of which he had taken possession of before the patent, but it did not avail, as he had to implore Lord Baltimore to save his life."

The ambassadors replied, "this was a different case; that they were not subjects of England, but of the Dutch nation, and had as much right to settle part of America as any others."

On the 18th, the governor again demanded to see their patent for South river.

The ambassadors replied they "had not had it with them, but they would show it at a future meeting."

There were also some remarks made on soldier hostilities, and that each must pursue his own course.

The ambassadors replied that they "should prepare themselves for defence," but at the same time solemnly protesting against such attacks. They also said they knew they (the Marylanders) would not attack them (the Dutch) in a clandestine manner.

To this the Council replied, "they would use their own pleasure. Payment for runaways," the Council informed them, "might be settled by their courts; but they could not compel them to return, because they considered Delaware in their jurisdiction."

At a further conversation between the governor and the ambassadors, the governor asked as to who was meant by the "Dutch Swedes" spoken of in their address.

To this the ambassadors made an answer that was hardly a truthful one. They replied "that they had been partners and associates, residing for some time (or rather connived at) under the jurisdiction of the Company, but they became so insolent, that in a traitorous manner they surprised New Amstel, then called Fort Cassimer, by which the Director General and Council of New Netherlands were compelled to cleanse that neighborhood of such a vile gang."

The Swedes did indeed attack and take Fort Cassimer, but they were never (until conquered by Stuyvessant) under Dutch jurisdiction, as the ambassadors intimated. Fendall, it appears, was ignorant of the settlement of the Swedes, and the conquering and occupation of their territory by the Dutch, although how this could be, when there were so many Swedish refugees in Maryland, it is hard to imagine.

Again, the ambassadors in their communication to the Marylanders, mistated the time of the destruction of Fort Oplandt and the murder of the first settlers of Delaware. They alleged it occurred previous to the building of Fort Nassau, in 1623. When this event took place, in 1631, or nine years later, and only one year previous to the grant of the State of Maryland to Baltimore, which was in 1632.[1]

Fendall also inquired of the ambassadors with great anxiety in relation to the "mountain" they had mentioned as a place of meeting, from which (as the ambassadors said), "the Sassafras river, in Virginia, and the kill which emptied itself into South river, behind Reedy Island, seem to derive their origin." They said, "we had our passage over this mountain." This, we think, must have been either Chestnut or Iron Hills, near Newark, which would have been in their course, more than probable the latter. They are situated in Pencader hundred, about two miles from the town of Newark, and about the same distance from the Maryland line. The Sassafras river and the Augustine creek, which flows into the Delaware back of Reedy Island, do not indeed take their origin there, as Herman and Waldron supposed, but Persimmon creek, which flows into the Christiana, and a branch of the Elk river, which takes its course through the

[1] Subsequent investigation has led the author to believe that the earlier historians who have written on the Delaware river, have been wrong in placing the scene of the massacre of the early settlers of Delaware at Lewistown. Paradise Point is laid down by Lindstrom as being on the southern bank of the Mordare Kÿlen, or Murderer's Creek. Lindstrom arrived only 12 years after this massacre, and therefore, whilst it was fresh in men's minds. It was at Swanendale the unfortunate settlers built their fort. Therefore there is but little doubt that on the southern bank of the Murderkill Fort Oplandt was built, and that that creek derives its name from the tragedy enacted on its banks.

Cat Swamp, have their rise within about half a mile of each other, near the Iron and the Chestnut Hills.

The ambassadors also had some private conversation with the governor on the subject of establishing mutual trade overland between Maryland and Delaware Bay.

Herman assured him "that this could easily be carried on, as soon as this question was terminated, and the limits of both sides adjusted."

This trade it was intended should have its course overland from the Bohemia river, in Maryland, to the Appoquinimink creek, in Delaware. It was intimated to Herman that this trade by land would be less likely to excite the jealousy of England than if it was conducted by sea. This hope of trade between these places was undoubtedly the reason that induced Herman to obtain the grant of the land now known as Bohemia Manor.[1]

On the 20th, the ambassadors departed. Waldron to the South river, with a relation of their proceedings, and Herman to Virginia, for the purpose, as he wrote, "of inquiring of the Governor of Virginia what was his opinion on the subject, and to create a division between them both (*i. e.* Maryland and Virginia), and to purge the Dutch of the slander of stirring up the Indians to murder at Accomac."

Thus ended the embassy of the Dutch to Maryland.

Herman in a letter to Stuyvessant recommended

[1] See Herman's account in Albany Records, vol. 18, pp. 337–364, and in vol. 2, Broadhead and O'Callighan, pp. 80–98.

the Board of Directors at New Amsterdam to appoint one of their number to visit Lord Baltimore "to see whether an agreement could not be made quietly." He also recommended the drawing of a corrrect map of the South river and Virginia, in which the lands and hills (creeks) should be laid down on an exact scale of latitude and longitude.[1] These things he wished done before complaints were made by Baltimore to the English government.

The Rev. Everardus Welius, the first clergyman we have a record of as residing at New Castle, died on the 9th of December.[2]

At this time the following mechanics were employed at New Amstel. They are the first named as following these trades in this State, viz.: Andries Andriessen, a carpenter; Theunis Servaes, of Harlem, a cooper; Cornelius Theunissen, a smith; William Van Raesenberg, a surgeon; Thys Jacobsen, a boy working at carpentering with Andries Andriessen; he is the first carpenter's apprentice recorded. There were also Joost, of Amsterdam, and Antony Willimsen, of Vreedlandt, masons.

In the meantime, the Dutch were still suspicious of the Swedes. Some concealed powder was discovered in a desk, and they were afraid that some one who was working mischief was concealed amongst them.

Beekman, the Governor north of the Christiana, became very sick, and things generally on the South river were in very bad condition.

[1] Broadhead and O'Callighan, vol. 2, pp. 99, 100.

[2] In 1657 or 1658, the Rev. John Polhemus organized a church at New Castle, whilst on his way to Brazil.

Stuyvesant wrote to the Directors in Holland, giving an account of the disputes with the Marylanders. He expressed the opinion that "they would take the first opportunity to expel" the Dutch from the South river. He urged the strengthening of that river, and at the same time informed them that "Governor of Maryland had already caused a survey to be made of the lands at the distance of about one or two miles[1] from Fortress New Amstel, and caused a distribution to be made of them amongst the inhabitants of Maryland. He desired information if they took possession of these lands, what should be his course of proceedings."[2]

The City of Amsterdam found that the Colony of New Amstel was of great expense, and no profit to them. On the 30th of September they appointed a committee of their Council to confer with the West India Company in relation to surrendering it back to them on equitable terms. On the 8th of November no agreement could be concluded to that effect.[3]

In addition to other troubles that afflicted the unfortunate Dutch colonists of the South river, were petty disputes between the officials of Altona and New Amstel. The officials of the Colony of the City believed that those of the Company were persuading the settlers to desert their territory and re-

[1] The reader must bear in mind that in Dutch correspondence, Dutch miles are meant, which are three English miles.

[2] Albany Records, vol. 10.

[3] Broadhead and O'Callighan, vol. 2, p. 111.

move to New Amsterdam. Several residents of New Amstel declared before their Council on the 14th of November that the officers of the Company, amongst them Cornelius Van Ryven and Martin Krygier, had held out inducements for them to desert New Amstel and remove to Manhattan. These declarations, signed by the parties, were sent to the Burgomasters of Amsterdam.[1] These disputes finally broke out in open quarrel between them in regard to a frivolous matter—the cleaning of the Fort at New Amstel. Captain Krygier ordered a sergeant to assist in this work, who was one of the city's soldiers. He refused to obey the order, stating he was forbidden by Alricks and D'Hinoyossa "to obey any other command than theirs." This, Krygier afterwards naively remarked in a letter to Stuyvesant, "sounded in our ears as an uncommon trumpet." Krygier, however, directly afterwards addressed himself to Alricks in presence of D'Hinoyossa, expressing his surprise at the command, telling him "that he must know" by his "credentials and instructions with what high commission he was endowed, and that he wished to know if it was done with his (Alricks) knowledge."

Both Alricks and D'Hinoyossa then declared "that the City of Amsterdam's servants were not holden agreable to their oath to obey any further commands than those of the city." And D'Hinoyossa further declared that "no one, while he held commission, should hold command over him, or the soldiers of the city," and such other discourses, which, says Krygier in

[1] Broadhead and O'Callighan, vol. 2, pp. 103–6.

his letter to Stuyvesant "should not be passed without protest, yet we do it, as it might lead to discussions, and to be avoided. We trace it, however, to the oath which had been taken, excluding the Directors of the West India Company."[1] Krygier proposed alterations in this respect.[2] Other matters of annoyance also took place between them. Van Ryven and Krygier on the side of the West India Company finding constant fault with Alricks and D'Hinoyossa, the officers of the City of Amsterdam, caused the latter to retort, and Alricks charged the company and its commissaries "with all the trouble that had been raised in the City's Colony by the desertion of its citizens and soldiers. In one of his letters Alricks alleged that if the "Colony of New Amstel, or any place depending on it, was lost or ruined, that they (Stuyvesant and the other company's officers) would be to blame for it." Alricks also protested against "recalling the garrison from the Hoernkill." Stuyvesant severely censured him for this. In a letter to the Company he denounced this latter charge in the protest of Alricks, of "absolutely commanding the recalling of the garrison from the Hoernkill as "impudent and false."

On the 30th of December, Alricks, after a sickness of several months, died, having first appointed Alexander D'Hinoyossa as his successor, and Gerritt Van Gezel, secretary.

Alricks' appointment was unfortunate for the

[1] Albany Records, vol. 18.
[2] Ibid, p. 234.

Colony, principally on account of his "too rigid preciseness," as his strictness in collecting debts due the city of Amsterdam (and his holding the unfortunate emigrants from Holland to their agreement to stay four years), was called by Stuyvesant. But it must not be forgotton that he had a difficult task to perform, and that the records from which we mainly derived our information as to his character were written by those opposed to him. And again, the city of Amsterdam dispatched over vessel after vessel loaded with colonists, who were not agriculturalists, or men fully master of any mechanical art useful in a new country. In many cases they were of sedentary occupations (such as weavers, tailors and buttonmakers) unaccustomed to hard bodily labor, who however valuable in an old and thickly settled community were useless in an uncleared wilderness, which was then the condition of what is now the State of Delaware. Again many of those sent over were vagabonds, without any legitimate occupation, who infested the streets of Amsterdam. With people of of this description, that city swarmed the colony of New Amstel, without sending sufficient food for their support.

The ship Meul, with one hundred souls, arrived at one time without a mouthful of provision; and then at another time the Indians destroyed their corn. In addition to other evils the settlers were attacked by the bilious fever, which disease is always prevalent in neighborhoods where wild lands are being cleared for cultivation. Delaware was no exception to this rule. Even in the memory of men still living this disease was very prevalent in what are now portions of the city of Wilmington; and in certain houses this malady annually entered, and struck down the inmates with sickness and death. The old Whitehall property still standing near Church and Ninth streets, which at the time of its erection was considered one of the handsomest residences in Delaware, was especially noted for its unhealthiness. It was then the mansion of a farm or plantation. At a residence situated at Seventh and French streets, which was then the summit of a hill, the bilious fever would annually enter and prostrate the household. There were other houses in Wilmington where it was almost impossible to live and retain health, and where almost annually one of the family would be carried to the grave. In portions of Kent and Sussex it was especially fatal. Children were raised with difficulty; and nearly every family would have to mourn the loss of portions of their younger offspring.

The clearing of the woods and draining the swamps and marshes has greatly ameliorated this disease in all portions of our State, and in some sections driven it away entirely. But it especially marred the efforts of Alricks, leaving him with a horde of sick and helpless people, whose energies were destroyed, with so little food to feed them, that it is alleged many of the unfortunate settlers of New Amstel died from starvation.

CHAPTER XXIII.

FROM 1660 TO 1661.

Dispute between D'Hinoyossa and Secretary Van Gezel—He flies to Altona to save himself from arrest—His removal from office—Appointment of Prato as Councellor—Van Sweringen as Secretary—Peter Alricks appointed commander at the Hoernkill—Orphan house at New Amstel—First Delaware orphan—Murders of Indians—They threaten revenge—Information sent to Stuyvesant—He urges punishment of the murderers—Sends commission to try them—Tried and sentenced previously by D'Hinoyossa—Payment made to the savages as a recompense—Robbery of Hudde—Fears of invasion from Maryland—Fort Christina decaying—Beekman attempts to move the Swedes—They refuse—Number of Swedes able to bear arms—Swedes receive permission to stay—Removal of Van Dycke the Swedish sheriff—Runaway Maryland servants delivered up—Horses on the Delaware—First divorce case—Criminal trial—Stuyvesant writes a letter censuring D'Hinoyossa—Indian sachem visits Beekman—Ferry at Hoernkill—Utie agrees to inform on runaways—Attorney for Baltimore demands the delivery up of New Amstel from West India Company—They refuse—They lay it before the State General, who instructs their ambassador to lay it before King Charles—Dispute between Beekman and D'Hinoyossa—D'Hinoyossa will not be commanded by Stuyvesant—City of Amsterdam confirm D'Hinoyossa as governor—His conduct approved of—The Rev. Mr. Laers—Augustin Herman—First roads in Delaware—Founding of Appoquonome (Odessa)—Amsterdam desires to give up New Amstel—Agree to hold it.

[1660] THE death of Alricks and the appointment of D'Hinoyossa as governor of New Amstel, it does not appear, worked more advantageously to the benefit of the settlers. The jealousy between the officers of the two colonies of Delaware (Altona and New Amstel) still continued. A dispute occurred between D'Hinoyossa and Van Gezel, who was a nephew of the deceased governor Alricks, in relation to the latter's estate, and Van Gezel had to fly to Altona to save himself from arrest by D'Hinoyossa, and request Beekman to protect him from his (Hinoyossa's) violence. Upon this D'Hinoyossa removed him from his office of councellor and secretary. John Prato he appointed to the former office, whilst the sheriff, Van Sweringen, acted as secretary. D'Hinoyossa also, upon Van Gezel refusing when summoned to appear before him, entered his house and took therefrom a mirror and picture valued at twenty-five guilders.

The government of New Amstel at this time consisted of D'Hinoyossa as governor, Van Sweringen and Prato as councellors, whilst they called to their aid on extraordinary occasions Williams, the surgeon, and John Block, the gunner. Peter Alricks was appointed as commander of the Hoernkill.

Beekman, on the 1st of February, received a note from D'Hinoyossa, without any address, making inquiries in relation to Van Gezel, and offering as an excuse for its want of direction, "*that he had no time to write the address without breaking in upon his laziness.*" Of this Beekman sent on a rather sneering account to Stuyvesant. D'Hinoyossa complained that Van Gezel had not rendered either his accounts as an auctioneer, or those of the *orphan house.* For at this early period had the Dutch an institution to provide for friendless children. The first child placed in this house was born on the Prince Maurice, wrecked on Long Island, whilst on her way to this country with settlers for New Amstel. Its father was named John Barneston. He was murdered by the Indians. Its mother died at Colonel Utie's. Its parents appear to have fled from New Amstel to Maryland. It was named by the Burgomasters of New Amstel "Amstel's Hope."

The first criminal trial we have any mention of now occurred in our State. Gerrit Herman and Govert Jansen having quarrelled, Jansen with his sword wounded Herman in the palm of the left hand, and cut off his finger; he was sentenced to pay the account of Herman, also his surgeon's bill, sixty guilders in money, and to work for six weeks at the spade and wheelbarrow in the fort at Altona. The sentence was signed by Beekman, and on the 31st of May approved by Stuyvesant.

In addition to internal trouble, and the uncertain state of affairs with Maryland, the unfortunate Dutch settlers were now in danger of a war with the Indians. Three Indians were found murdered on the farm of Jacob Alricks, the deceased governor, near New Amstel, on the 21st of January. One of these was a Minqua, or one of those Indians who resided on the Christina river. The murder, it was alleged, was committed by two of Alrick's servants. The bodies were found in the underwood, in the marsh, by some Indians, who communicated the information of the murder to their tribe, who at once threatened to take revenge on the residents of New Amstel. The neighboring inhabitants, upon this, abandoned their residences, and fled to the fort for protection. Much indignation was excited against D'Hinoyossa on this occasion, on account of his saying that "he would not contribute a farthing in the case of this murder, but that it must be borne by the community, and that he was "pretty indifferent whether the savages went to war or not." Beekman, however, endeavored to settle the matter peaceably with them, and sent for Van Dyke, the Swedish sheriff, to consult with the authorities of Altona and New Amstel, to devise means to prevent the threatened bloodshed.[1] The supposed murderers were apprehended, and information sent to Stuyvesant, giving him a full account of the affair. He wrote back, urging the importance of the conviction and execution of the murderers. As he could not go himself to the South river, he sent his attorney-general, Nicatius de Stille, and Paulus Lindert Van de Graft, an old burgomaster of Amsterdam, who, with Beekman, D'Hinoyossa, Van Sweringen, Jacobus Backer, (acting schoepen,) and John Prato, were to inquire into the circumstances of the murder. They were instructed as follows:

"When the inquiry is made, the delinquents discovered, and by sufficient proofs and voluntary confessions convicted, then prosecute them before the delegated Judge to make up his conclusion according to law, demand speedy and impartial justice, and execute the pronounced judgment there on the spot for others' example.

[1] Albany Records, vol. 7, p. 29.

"Shall invite the sachem and some other individuals to be present, and explain it as an object of friendship, and that they may be made to do so when Indians kill whites."

They also brought with them some other instructions, viz., to inquire into the case of a man named Becker, for selling rum to the soldiers, some of whom, whilst in a state of intoxication, had burnt a canoe belonging to the Indians. Becker was tried, convicted, and dismissed from his office of clerk for this offence. They were also to exhort D'Hinoyossa and Alricks' executors to peace, and to advise and assist Sergeant Andreas Laurens in the military service, whom they authorized to "enlist Swedes and Fins as soldiers at eight or twelve heavy guilders per month."[1]

Nicatius de Stille and Van de Graft arrived at New Amstel on the 8th of March, for the purpose of composing part of the court to try the murderers of the Indians, but on the 10th of February, nearly a month previous, D'Hinoyossa had tried, convicted, and sentenced the murderers on his own responsibility. He solicited the attendance of Beekman at the trial, who at first declined to be present, but afterwards attended. Beekman asked him if he "supposed himself sufficiently qualified to decide such cases." D'Hinoyossa answered "Yes." He then requested Beekman to "take a seat near the fire and hear the debate and decision." To this Beekman consented. The alleged murderers were convicted, but they appealed from

[1] Albany Records, vol. 24, pp. 108, 109.

the judgment of D'Hinoyossa to the director and council of New Netherlands at Manhattan, by whom it appears that the judgment of D'Hinoyossa was reversed. This was the first trial for murder that took place in Delaware.

On the 18th of August, payment was made by Beekman to the savages as a satisfaction for the murdered men, and a receipt signed by them given to him. Whilst this was being done, a band of Indians attacked and robbed Andreas Hudde, formerly director or governor of the Dutch possessions on the Delaware, at the time they were mainly in and around Fort Nassau, and on the site of the present city of Philadelphia. The Indian sachems, upon being informed of this outrage, engaged that every thing should be returned. This, however, was never done, and poor Hudde, who had proved himself a faithful servant of the Dutch, was reduced to poverty.

The Dutch were still alarmed by rumors that Baltimore would invade the South river. On one occasion, fearing an attack from the Marylanders, they pulled down an old house on Cooper's Island (which was situated on Cherry Island Marsh, now within the corporate limits of the city of Wilmington) to get timbers to repair Fort Christina, which was in a decayed condition.

Attempts were still made by Beekman, in accordance with the instructions from Holland to get the Swedes to change their habitations, but strong objections were made by them to moving from their settled and cleared lands, where they had already erected their dwellings, to others, wild and uncultivated, and destitute of buildings. The Swedes and Finns appear to have been much annoyed at this constant interference with their habitations, and some twenty families prepared to leave the company's colony of Altona and reside in the city's colony in New Amstel.

At this time, according to the report of Van Dyke, the Swedish sheriff, to Beekman, there were one hundred and fifty of the Swedes capable of bearing arms.[1] The Swedes and Finns were two separate people, and could not converse with each other on account of the difference in their language. Originally, the land between Marcus Hook and Chester, Pa., was called Finland, and here undoubtedly was the principal settlement of that people. Endeavors were made to get them to settle at Passyunk, a territory lying between the Wicaco and the Schuylkill, situated within the present limits of the city of Philadelphia, but they declined on one pretext and another. First soliciting a delay until after harvest, and at last peremptorily refusing to go. Attempts were then made to get them to settle at Esopus, now called Kingston, in the present State of New York, but as the Dutch were engaged in war with the Indians in that neighborhood, they very properly declined to reside in a vicinity which was then the scene of massacre and murder. Eleven Indians at Esopus had just been slaughtered by the Dutch, and the whole of the Indian tribes there were banded together to revenge

[1] Beekman's Letters, vol. 17, p. 45 of Albany Records.

the murder. Beekman thought "they were admonished and encouraged by some of the principal leaders among them not to disperse, but to remain on the South river as closely united together as possible." They finally received permission from Stuyvesant to stay, but not until they informed Beekman that if compelled to go, "*they would depart to a spot where they might live in peace.*" Van Dyke was afterwards discharged from his office of sheriff, on the ground that he had influenced the Swedes against moving from their settled habitations to other places marked out for them by the Dutch.

On the 2d of April, a Mr. Henry Coursay, a merchant of Maryland, arrived at New Amstel, seeking for some runaway servants of his. D'Hinoyossa at first refused to deliver them up, but finally agreed to do so, and they were given to him at the Hoernkill on the 11th of April. Beekman thereupon sent to the governor of Maryland, to Colonel Utie and the magistrates residing on the Sassafras river, a complimentary letter, in which he "requested that if any Dutch soldiers deserted to Maryland, that he would arrest and imprison them, and inform him of it by express, at the expense of the company, in which case he would despatch with their permission, a sergeant, assisted by soldiers, to accompany them home."[1]

We have an account of the number of horses on the Delaware at this time. They appear to have increased but slowly. Beekman notifies Stuyvesant that the "horses are misused by the Swedes," so that he feared

[1] Albany Records, vol. 17, p. 49.

there would be "no increase by them." He stated "that the mares were spoiled by drawing the whole morning heavy beams," and that "only three mares and two stallions were alive" of the number Stuyvesant sent, "besides two young colts of two years."[1]

We have in the dispatches of Beekman this year some accounts of criminal trials, and also glimpses of the darker shades of social life in our State at that time. He says:

"Amongst the Finns is a married couple who live together in constant strife. The wife receives daily a severe drubbing, and is expelled from the house as a dog. This treatment she suffered a number of years; not a word is said in blame of the wife, whereas he, on the contrary, is an adulterer. On all which the priest, the neighbors, the sheriff, appeal to me, at the solicitation of man and wife, that a divorce might take place, and the small property and stock be divided between them." Beekman asked for orders. This is the first mention made of any divorce case in Delaware.

Beekman also gives an account of two other trials, one for marrying illegally, the other for assault and battery. In the one case, Oloff Stille, an influential Swede, was a prominent actor. He was the resident of a village named after him by the Swedish freemen "Stillensland," situated on the Delaware, not far from Chester, Pa. Stille had a thick black beard, from which the Indians gave him the name of "the man

[1] Albany Records, vol. 18, pp. 51, 58.

with the black beard." He was the ancestor of the respectable Stille family now residents of Philadelphia. Beekman, in giving his account of the trial,[1] which took place at Altona, says: "Oloff Stille opposed himself to me pretty warmly in court, because I suspected him, that he, without being authorized, had arrogated to himself to qualify the priest to marry a young couple without the *usual proclamations*, and against the will of the parents, on which I condemned the priest in a fine of *fifty guilders*, which said Stille too opposed, saying that it was not our province to meddle with this affair, it ought to be done, if any interference was desirable, by the Swedish Consistory, and that we had nothing to do with the priest. Mr. Laerson[2] adopted the same opinion, as our court related to us, last November, on a summons, that we had no right to interfere with the rights of Christina, so that he did not appear before us. The case was this: Mr. Laerson had complained of assault and battery by Peter Mayer; he was severely struck and wounded in his face, so that I never saw a worse, on which both were summoned to appear before us; and before the court could meet, the affair was settled, pretending the incompetency of the court. On the 19th of August the court met. There were twelve Swedish and Finnish nations summoned, and a default by the Jagers and other Dutchmen in the colony,

[1] Campanius.

[2] Laerson was the Swedish minister left by the Dutch, and spoken of as of "godless and scandalous life." He is named by some as "Laurence Charles Lokenius." He is also called "Laers."

when an order was issued that for each default, which was voluntarily and premeditated, and not brought forward by any extraneous or invisible obstacles, as through sickness, or on God's wind and weather, should be paid a fine of ten guilders, so that no person should be delayed in his just pretensions, as there were annually only three or four courts, as circumstances might require. On the 7th of this month, being court, the priest and Mayer were again summoned on the same affair by Sheriff Van Dyke; and, further, that Peter Mayer treated another person in the insolent manner, and Peter Mayer, deliberately occasioning default, after the eight intimations, to Jacob Van Vern, in whose behalf the assignment was made by the sheriff and commissaries for liquor received; on which Peter Mayer, coming to me, requested a receipt, making at the same time a great noise, that in this manner the costs were excessive, but that he too would not be at rest until he had legal security for his land. He told me, further, in pretty harsh language, that every year new commissaries ought to be appointed, as entitled, or that he and other freemen were always to be treated as boys, so that constantly we are to be ruled by madcaps, who did not understand reading or writing, and were to be preferred before him, who was acquainted with letters and penmanship, and that the affairs should be managed in quite different manner, if he should remain here; with many other similar insolent blubberings; to all which I listened with patience, refuting him with solid arguments, and advised him to go to your honor and lay his complaints before you. A few days ago, when I sent him warning to deliver up his horse, his wife came here and made a horrible noise; they could not spare the horse; they were not accustomed to carry their wood on their necks, and they had a share in the property of the horse as well as I; and, be it said with reverence, she did not care a groat about my orders, as they intended to leave soon this spot; on which I menaced to send her to the guard house; but having no wish to throw her in consternation, as being in her last stage of pregnancy, I let it pass by. In short, this people conduct themselves most despicably. Nevertheless, if they are resolved to move from here and reside in the colony, or any other part, I shall remind him of the fine which he yet owes for selling liquor to the savages. I am informed that the greater part of those now living separately do intend going to reside in Maryland with a few of the Finns."[1]

Stuyvesant appears to have severely censured the conduct of D'Hinoyossa, the governor of New Amstel, as Beekman, in a letter to him of April 8th, says:

"I copied your letter to the Honorable D'Hinoyossa, sealed it, and then directly conveyed it to him. He was, he said, much surprised at such a vulgar address, covering such bitter contents, and you might be assured that he would show it from the greatest to the smallest, and even forward it to the States. He said, further, he would not do any thing contrary to your orders, but bear all things with

[1] Albany Records, vol. 17, pp. 51, 58.

patience, but remonstrated against it to his principals. He wished to be informed by me if the colony was again transferred to the company. He understood at least by your letter that a change was in contemplation."[1] Beekman also complained of D'Hinoyossa not restricting the sale of liquors to the Indians in daylight.

On the 23d of May, the great chief of the Minquas visited Beekman, and showed him his poor coat. The hint was taken by Beekman, who thereupon presented him with a coat and piece of frieze in the name of Stuyvesant. On the 17th of June, a sachem arrived at Altona from Hackensack, with three or four other savages, among whom, it was said, was the brother of a sachem who was killed at Esopus by the Dutch. Accompanying these savages was the great chiefs of the Minquas, who informed Beekman that he intended visiting New Amsterdam the next night to see if he could make peace between the savages and the Dutch. As to whether the great chief of the Minquas visited Stuyvesant on that peaceful mission, or as to the result of that mission if he did, history is silent.

There appears to have been a ferry at the Hoernkill at this early day, as the records say the boat used for it was sunk. By the same record we learn Colonel Utie agreed to inform the Dutch of the runaways from the Delaware to Maryland.[2]

On the 1st of September, Captain Neal, the attor-

[1] Albany Records, vol. 17, pp. 51, 58.

[2] Albany Records, vol. 18, p. 80.

ney of Lord Baltimore, in obedience to instructions from him, had an interview with the College of 19 in Amsterdam, and asked the Directors (much to their surprise) to deliver up the Dutch Settlements on the Delaware, in the neighborhood of New Amstel and Altona to him. This claim was made (Neal informed them) under the grant from Charles the 1st. If the demand of Neal was complied with, Baltimore agreed to give them indemnity for "*all costs, damages and interests already undergone or to be yet incurred.*" They replied by asserting their "right by possession, under the grant of the State General for many years, without disturbance from Lord Baltimore, or any other person." They declared they were "resolved to remain in possession and defend their rights," and "if Lord Baltimore persevered and resorted to violent measures, they would use all the means God and nature had given them to protect the inhabitants," and "would be innocent of any blood which may be shed."[1] The College (or West India Company, by which name it is better known to our readers) on the 5th of November following, laid the matter before the State General, and requested them to "represent, through their ambassadors in England, the situation of affairs with Maryland, and to complain of Lord Baltimore's encroachments and pretensions," and to desire that the English government should require that Baltimore should "desist from them until a boundary line could be run between the provinces of Maryland and the Dutch." The State General agreed

[1] Albany Records, vol. 8, pp. 294, 296, 301.

to the request of the Company, "and all necessary papers were furnished to the ambassadors, who were instructed to direct the attention of King Charles the 2d to the subject.[1]

The dispute between the jurisdiction of the company and the city appear to have continued through the whole of 1660, and Beekman made complaint of D'Hinoyossa in every letter he wrote to Stuyvesant. He complained of him for allowing liquor to be sold "to the savages, so that they behave shamefully;" also "of his administration of the estate of Alricks." In one of his letters he says that he (D'Hinoyossa) says that "he will not be commanded by your honor (Stuyvesant), as he does not acknowledge any person his superior, except his principals in fatherland."[2] Again he wrote, (speaking of his administration on Alricks' estate,) "D'Hinoyossa is mentioned as conducting himself in a haughty and imperious manner, defaming and slandering the deceased director (Alricks), and disregarding mandamuses, and injuring the property of the deceased."[3] But the conduct of D'Hinoyossa appears to have met the approbation of the city of Amsterdam, as on the 27th of August the magistrates of Amsterdam appointed him as director, and John Prato and Gerritt Van Sweringen as assistants. They also resolved to continue their support to New Amstel, and approved of his conduct in seizing the property of Alricks.

[1] Holland Documents, vol. 9, p. 144; O'Call. vol. 2, p. 461.
[2] Albany Records, vol. 17, p. 92.
[3] Albany Records, vol. 18, p. 141.

This intelligence arrived in New Amstel by way of Maryland, on the 24th of December. Beekman, in describing the arrival of the intelligence, says: "In short, the joy is here great, so that the seal of the letter was scarcely broken, than he commanded the cannon to be three times fired."

After the death of the Rev. Mr. Welius, the only clergyman on the South river was the Rev. Mr. Laers, Lokenius, or Laerson, (for he is known by all three names). The Swedish or rather Finnish minister (for he was a native of the latter country), and his character was such that he could not command the respect of the Dutch. Four clergymen were expected to arrive at New Amsterdam from Holland. Beekman petitioned that one might be sent to Altona.

Augustine Herman appeared to have remained and settled in Maryland, and at this time obtained the grant of Bohemia Manor. This grant was made to him on account of his making a complete map of Maryland and Virginia, which he dedicated to Charles the 2d, King of England. In the Greenville Library is a copy of this map. It was made by Fairthorne, an artist distinguished for crayon portraits and copperplate engraving. On it is this statement, "Virginia and Maryland, as it is planted and inhabited this present year, 1670, surveyed and drawn by Augustus Hermann Bohemiensis." This map also contains a beautiful portrait of Hermann.[1] He was also instrumental in constructing a cart road which was made

[1] See Neill's Terra Mariæ.

about this time between the Bohemia river, in Maryland, and the Appoquinimink creek, in New Castle county, Delaware. The making of this road caused the building a village on the Appoquinimink creek, called Appoquoneme.[1] This village was the ancestor of the present thriving town of Odessa, formerly known as Cantwell's Bridge. Odessa is therefore the next town in our State in antiquity after Wilmington, New Castle, and Lewistown. This was the first road we have any account of, as being constructed in this State. He also endeavored to establish a village on the Bohemia river. Writing to Beekman, some time later, he says: "I am now engaged in encouraging settlers to unite together in a village of which I understand a beginning will be made before next winter. From there we may arrive by land in one day at San Hoeck (New Castle). * * * The Maquas (Christina) kill, and the Bohemia river are only one mile[2] distance from each other, by which it is an easy correspondence by water, which would be greatly encouraging to the inhabitants of New Netherlands."[3] This village is now called Port Herman, but it has dwindled to nothing. It has not thrived like its Delaware sister. Herman appears to have made successful efforts in softening the differences between the English and the Dutch, and rendering the intercourse between them pleasant. It was undoubtedly owing to his influence that the fierce Marylanders

[1] Journal of Dankers and Stuyter. See Memoirs of the Long Island Historical Society, vol. 1.

[2] A Dutch mile is three English miles.

[3] Vol. 17, p. 111.

were modified and induced (instead of marching with armed forces against the Hollanders) to open a trade with them that would tend to their mutual advantage. He wrote several letters to Beekman, advising him in relation to matters occurring between the citizens of the South river and Maryland, and appeared deservedly to have acquired the confidence of both Dutch and Marylanders.

The colony of New Amstel had proved an expense to the city of Amsterdam instead of a profit, and they desired to convey it back to the company. They appointed a commission to see if arrangements could not be made for that purpose on the 30th of September, 1659, but no satisfactory terms could be agreed upon between them. To the 1st of January it had cost them 165,200 guilders, for which outlay there had been little return, and they had been dunned for the interest of the loan. On the 16th of March (as they could not get rid of it) their Council passed a resolution to retain the colony. The commission they had appointed to inquire into its affairs, reported to them that amongst other causes tending to injure New Amstel was the interference of the company's officers with those of the city's colony. They were of opinion that this difference about jurisdiction might be remedied by the "company holding their director (Stuyvesant) to his duty, and *sharply* interdicting him from undertaking any thing contrary to the right of the city's colony." The commissioners also reported that the right of appeal to the Director General and Council of New Netherlands, in cases where the sums in dispute were over one hundred guilders, and of appeal in criminal matters, and the claiming of dues for anchoring in front of New Amstel, and compelling vessels carrying goods for the South river to unload at New Amsterdam, were amongst the other things detrimental to New Amstel. The Council of Amsterdam accordingly conferred with the West India Company, who agreed that the courts of New Amstel should have jurisdiction in actions to the amount of 600 guilders; that there should be no appeal in criminal cases; that the city of Amsterdam should have the appointment of sheriff; and that vessels should be allowed to proceed direct to New Amstel and discharge their cargoes, without touching at New Amsterdam. There were also other minor considerations and alterations made in the original compact between the company and the city. The city of Amsterdam[1] also issued proposals, inviting merchants and others to engage with the city in the planting and trade of New Amstel. The city had hopes of an increased prosperity of their colony from the opening of a trade with Maryland. They speak of a certain creek (in the report of their commissioners) that has been discovered, that is navigable with small boats, within a quarter of a mile of the English creek. The creek alluded to was undoubtedly the Appoquinimink.

D'Hinoyossa had, in a letter written at New Amstel the 12th of December, 1659, and received by the city commissioners the early part of this year,

[1] Broadhead and O'Calligham, vol. 2, p. 172.

offered, if the city of Amsterdam advanced one hundred thousand guilders, to plant a thousand souls on the South river, and give them security for the money. He blamed the failure of the colony to Alricks. He accused him of "getting all he could gripe and catch, provided only it could be got on credit." This letter was written a few days before Alricks' death, who believed to the last that D'Hinoyossa was friendly to him. It undoubtedly encouraged the city of Amsterdam to persevere in the establishment of their colony, and caused them to believe that better management of the affairs of New Amstel would make it a profit to them, instead of a loss, which it had heretofore been.

CHAPTER XXIV.

FROM 1661 TO 1662.

Murder of three Englishmen and a Dutchman by Indians—Arrest of savages with their clothes in New Amstel—They are let go—Indignation of the Marylanders thereat—English suspect the Dutch of inciting the Indians to murder—Indians alarmed at the English—Hold a meeting at Passyunk — Present seawant to Minquas to reconcile them—Minquas present furs to Marylanders for that purpose—They refuse—Desire Minquas to form alliance to destroy the Indian murderers — Minquas refuse—Peace made between Marylanders and English near Appoquinimink—Dutch to supply Marylanders with negroes for tobacco—Grave of an Indian sachem violated and robbed—Maqua and Seneca Indians at war—Marylanders assist the former—Senecas destroy plantations in Maryland—Effect on Swedes and Finns—Catholics visit New Amstel and Altona—Wish to settle—Not encouraged—Unpleasant conversation between them and Beekman—Maryland Council decide not to press the rights of Maryland to New Amstel—Do not know whether it is in latitude 40—Wait for the will of the proprietary—Send agent to lay claim for the West India Company to South river—West India Company agree to give up Hoernkill—Stuyvesant censured in relation to New Amstel—His defence—First elopement in Delaware—Laers marries himself—Continued ill feeling between Beekman and D'Hinoyossa—Beekman's letters on the subject—D'Hinoyossa makes vessels lower their colors passing New Amstel—Refuses to see Beekman.

[1661] On the 4th of January, some Indians who resided on the Delaware murdered four people who were travelling from New Amstel to Altona. They were three Englishmen and a Dutchman (the brother of a Dr. Herck) who had been at Altona only a few days on a visit. Two or three days after the committal of the murder, several savages arrived at New Amstel, who had with them some of the clothes of the murdered men, which they offered for sale. It could not be ascertained whether they were the murderers, but two of them going to the house of a man named Foppo Jarison, were at once seized by him, assisted by William Hollingsworth, an Englishman, and Gerrit Ruster, an inhabitant of New Amstel. They immediately informed D'Hinoyossa of the seizure, who placed them in prison, the savages at the same time "uttering violent threatenings" against the inhabitants of New Amstel. After an examination they were set at liberty. The setting of these savages at liberty was very displeasing to Philip Calvert, who had succeeded Fendall as governor of Maryland, (and who was an illegitimate son of George Calvert, the first Lord Baltimore, by a lady who had accompanied him on a visit to Newfoundland and Virginia,) and excited suspicion in the minds of the English, so much so, that Augustine Herman wrote in private to D'Hinoyossa "that the English foster the opinion that the inhabitants of New Amstel and the Hoernkill secretly instigate the river savages to such misdeeds, which," says Beekman, in a letter to Stuyvesant, giving an account of the affair, "is certainly an odious and wrong imagination."[1] Herman also wrote, a short time afterwards, to Beekman, in which he said: "It is much disliked, and had given offence that the apprehended Indian murderers, who murdered the

[1] Albany Records, vol. 17, p. 109.

English, and carried their clothes to Sand Hoeck, (New Amstel,) have been set at liberty. The English require satisfaction from the murderers, or war, whichever they choose, on this point they are now quarelling with the Susquehannocks. It was desirable that D'Hinoyossa would write the governor in a manner to remove the broil, and conciliate his good will, as the English are inclined to live with us in harmony and friendship, and cultivate and promote commerce between the two nations, which they would not interrupt, unless compelled by force or necessity. I hope for a favorable answer. It is said here that the Susquehannock chiefs have been summoned to Sand Hoeck, and there is some suspicion that it relates to the aforesaid business."[1]

The energy and determination of the English of Maryland in their demand for satisfaction from, or war with the savages for this murder greatly alarmed the latter. Accordingly, some time in the latter end of May, they held a meeting at Passyunk, and collected a large portion of seawant to make presents of it to the Minquas to reconcile them to the English for these murders. The Minquas, had, some days before, endeavored to present to Calvert some furs, which he refused to accept, but requested them rather to unite with him and destroy the nation of the savages that committed the murder. This the Minquas declined to do. The Dutch at this juncture endeavored to reconcile matters between them, and to get the Indians to send ambassadors to the Marylanders.

[1] Albany Records, vol. 17, p. 114.

This they refused to do, being afraid to trust themselves in their power. They said to the Dutch, "the English killed some of us, and we killed some of them; the one pays for the other." This answer, however, did not satisfy the Marylanders, but finally, on the 6th of September, the Dutch succeeded in getting two chiefs of the savages residing on the South river to meet Calvert, in company with D'Hinoyossa and Peter Alricks, to negotiate a peace.[1] This meeting, at which Calvert was attended by his council, took place at Colonel Utie's.[2] This, however, was only a preliminary negotiation, as another conference was held in the month of October, between the head of the Appoquinimink creek (then called the Appoquinime) and the head of another creek that flowed into the Chester river. Only one savage chief was present on this occasion, and he was from the eastern side of the Delaware. A treaty of peace was here made between the Marylanders and the Indians, a solitary sachem acting for the latter; and, as the Dutch chronicle remarks, "had a joyful intercourse between them and D'Hinoyossa." At this meeting the Marylanders offered to deliver annually to the Dutch two or three thousand hogsheads of tobacco, either at the Sassafras river or Appoquinimink creek, provided the Dutch could supply them with *negroes* and other commodities. Beekman,[3] in his letter to Stuyvesant, giving an account of the troubles arising from this murder by the Indians, gave utterance to the following pre-

[1] Albany Records, vol. 17, p. 124. [2] Ibid, 137.

[3] Beekman's Letters, Albany Records, vol. 17, p. 142.

diction, which in part was soon to be verified. He said: "It is my humble opinion that if the English enter into a war with the savages, it cannot be but to injure the public welfare, and that the savages will again claim and take possession of all the lands, or *that these will be eventually settled with the English and Swedes*."[1] This accounts for the anxiety of Beekman to preserve peace between the Marylanders and the Indians. The Dutch had previously, in the beginning of January, fears of an attack from the Indians themselves. The grave of Hoppionewick, an Indian sachem who was buried opposite the house of Captain Krieger, had been violated, and some seawant, and three or four pieces of frieze, and some other things that was buried with him had been stolen. At this the savages murmured, and mischief was apprehended at New Amstel.[2]

The Maqua and Seneca Indians were at this time at war. The Marylanders assisted the Maquas with fifty men in their fort.[3] The Senecas, in retaliation, it is asserted, destroyed several plantations of the Swedes and Finns who settled in Maryland. This was more than probably the case, as Oloff Stille, who left the Delaware with some Finns, to settle on the Sassafras river, in Maryland, (where several of their countrymen, who had also left the South river, had before settled,) returned, and continued to reside on the South river. They found the position of their countrymen not as comfortable as had been reported.

[1] Beekman's Letters, Albany Records, vol. 17, p. 109.

[2] Albany Records, vol. 17, p. 96. [3] Ibid, p. 118.

The return of these Finns was more than probable the result of actual or threatened raids from the savages.[1]

The trouble and strife that at this time occurred between the Catholics and Protestants in Maryland, caused some of the Marylanders of the former faith to visit the South river, to see if they could settle under Dutch jurisdiction, without being delivered up to Maryland authority. On the 4th of February, Captain Wheeler and Ulrick Anthony, two Catholics from Maryland, visited both New Amstel and Altona, and proposed the question to Beekman, whether "if any resident of Maryland came to the South river the Dutch would give him up when demanded? or if it was presumptive that he would defend them?" Beekman pretended that "he was not obliged to answer, as, not being deeply enough versed in law, even if it was brought before him." He told them it was a question too critical for him to decide; the more so as Maryland appeared deeply interested in it, as being in duty bound to defend the privileges of their citizens." They appeared to be mortified at Beekman's answer, and excused themselves for their arrival and departure from his house by saying that they well knew how to show due respect, but in this case their conscience would not permit it." Beekman answered that his "conscience did not tolerate such a sect." In his letter to Stuyvesant relating, the meeting, Beekman says: "If they remain quiet, and no others of the same creed shall arrive hither, I

[1] Albany Records, vol. 17, p. 104, 105.

shall tolerate them till I have received your honor's further orders; but by an unexpected increase I shall command them to depart, in conformity to the orders of your honor." In a former part of the same letter he said: "Just as we were desirous to avoid making any encroachments upon the rights of our neighbors, it seems to me, in this question, with regard to us, every regard due to an independent State is lost."[1]

Some time in the month of May, the claim of Maryland to the Dutch possessions was again brought before the Maryland Council, when it was resolved that as it was a matter of doubt whether New Amstel lay below the 40th degree of north latitude, and as the West India Company resolved to maintain their possessions by force, and there was no prospect of any aid from the other colonies, in any attempt which they might make to reduce them, all further efforts for their subjugation should be delayed until the will of the proprietary could be ascertained, and that in the meantime some effort should be made to determine whether the settlement was located within the limits of Baltimore's grant. An agent was also despatched to Holland to enforce upon the West India Company the claims of the proprietary to the territory in question, and to repeat the demand that it should be abandoned. Compliance with this demand was again refused by the company, but it gave orders to its settlers to withdraw from the terrritory about Cape Henlopen, which they had purchased from the Indians.

[1] Albany Records, vol. 17, pp. 104, 105.

But New Amstel and the adjoining territory were still held in possession by the Dutch.[1]

The squabbles between the two settlements, the Company's Colony of Altona and the City of Amsterdam's Colony of New Amstel—accounts of which were undoubtedly sent to his employers (the City Government of Amsterdam) by D'Hinoyossa—caused the censure of Stuyvesant by the West India Company in Holland. D'Hinoyossa appears to have complained of the interference in the City Colony's affairs by the authorities of the Company, and the reversal of his judgments in the cases of the murderers of the Indians. In his letter of defence, written in July, Stuyvesant says:

"If all our letters written since the death of Alricks—written to his successor, D'Hinoyossa, and one or two of his principals—had been transmitted, it would, to all appearance, become evident that we instituted no process whatever against the City's ministers, much less that we sowed the seeds of discord amongst them; to the contrary, it would evidently appear that we most sincerely recommended them peace and harmony, with the promotion of the common welfare. What regards the case or process mentioned by your honor relative to the estate of deceased Alricks; our opposition against judgment on the murderers of three savages, in a case of appeal, and our unwillingness to the judgment against Jan Garrettson and Van Marcker, it was our duty to inform you of

[1] See McMahon, p. 25, who quotes Council Proceedings of Maryland.

both, and all the circumstances of these affairs, and our own honor and character compels us imperiously to express our minds with all possible brevity."

Then alluding to transactions connected with the estate of Alricks, and the petition of Van Gezel, Alricks declared heir and executor, he says: "Never was there an infraction upon arrest made by Provisional Director D'Hinoyossa in behalf of his principals or the estate, or money coming from it, or contracted for, which aforesaid remains responsible."[1] In this letter he speaks of Alricks as a man of discreet character.

The inhabitants of the South river were at this time greatly excited by a case of elopement—the first case chronicled as taking place within the State. On the 20th of September the wife of Mr. Laers, the Finnish priest, eloped with a man named Jacob Jongh. They departed in the night in a canoe, accompanied by an Indian. Messengers were despatched by Beekman to Sassafras river and other parts of Maryland, with directions to arrest the fugitives, but without success. Laers the night following went to the house of Andreas Hendriessen, a Finn, where Jongh resided, and broke open the door of Jongh's room, and also broke open his trunk, which was there. Laers, however, does not appear to have been inconsolable, or to have taken his loss much at heart, as in less than a month after (on the 15th of October) he solicited Beekman for his consent to marry again—to a girl seventeen or eighteen years old. He wished to make

[1] Albany Records, vol. 18, p. 143.

the proclamation on the 16th of October. Beekman delayed to give an answer until he should obtain Stuyvesant's approbation. On the 18th of November he again solicited Beekman's permission to get married, as "the situation of his family," he said, "imperiously required it."[1] On the 15th of December he solicited from the Council a divorce from his wife, and obtained it. He did not get married, however, until the 31st of January following, when he married himself—"a transaction," says Beekman in his letter to Stuyvesant, "in my opinion, under execution, entirely unlawful, and expect your honor's orders how to conduct myself in it."

Laers afterwards got himself in trouble by this transaction. Jongh, it is supposed, went to New England, as his trunk, with several articles of Mr. Laers' property, was afterwards found at Upland (now Chester).

The ill feeling between Beekman and D'Hinoyossa still continued, and almost every letter of Beekman's contained complaint of his conduct. In his letter of January 14, in describing D'Hinoyossa's carriage upon his late appointment as Director or Governor of New Amstel, he says: "He feels himself again pretty high, and is strutting forward in full pride. He is boasting that he will recover all the effects of the deceased Alricks, and sings already another tune. He removed from office his secretary, Van Nas, because he did not flatter his whims in writing the records."[2] On the 6th of September Beekman went to New Amstel to

[1] Albany Records, vol. 17, pp. 142, 144. [2] Ibid, p. 96.

converse with him in relation to his making vessels lower their colors as they sailed past Fort New Amstel, when he was stopped at the door by Van Sweringen, who was ordered to prevent his entrance, although D'Hinoyossa was notified by Alricks of his arrival. A small boy afterwards brought word to Beekman from D'Hinoyossa that "he could not speak with him with a good conscience, and advised him to go home."[1]

[1] Albany Records, vol. 17, p. 135.

CHAPTER XXV.

FROM 1662 TO 1664.

War between Minquas and Senecas—Dullness of trade thereupon—Trial of Laers, the preacher, for breaking open Jongh's room and marrying himself—His marriage declared illegal—He appeals to the Court—First mill near Wilmington—Grant of land at Hoernkill to Mennonists—New Amstel fears an attack from Indians—Alarm thereat—Indian murders—Friendly feeling of the Minquas—Monopoly of trade of Hoernkill granted to Alricks—Selling liquor to Indians forbidden—Penalty—Disputes between Beekman and D'Hinoyossa—Charges by the former against the latter—D'Hinoyossa abuses Stuyvesant and the Manhattans—Sells guns to savages—Other things to Maryland—Does not keep a day appointed by Stuyvesant of fasting and prayer—Chooses one of his own—Desires no more orders from Manhattan—Is censured by Stuyvesant—Hanging of a runaway servant—Meeting between D'Hinoyossa and Governor of Maryland—Beekman continues to complain of D'Hinoyossa—First protested bill of exchange—Death of Pappegoya—Small-pox among the savages—First Frenchman in Delaware—The City of Amsterdam complain of their laborers being enticed from New Amstel by the Company's Colony—First land grant in Wilmington—Victory of the Minquas over the Senecas—Pardon of Van Sweringen for killing a soldier—Marylanders march on Hoernkill—Dutch abandon it—Duties levied at Hoernkill—Baltimore invites settlers there—He visits New Amstel—Arrival of colonists from Holland—Fort Altona decaying—City of Amsterdam requests that Altona and the country above Christina Creek should be ceded to them—The Company decline the conditions—Offer to cede Fort Christina—They finally agree—The deed of transfer to City of Amsterdam—Laborers escaping from City's Colony to Manhattan to be given up—City to send 400 colonists over—Farms, cattle, sheep and swine on South river—Death of Hudde.

THE war still continued to wage between the Senecas and the Minquas, which caused [1662] a stagnation of trade on the South river, as the Indians, whilst engaged in slaughtering each other, neglected their hunting, and therefore the procurement of furs.

On the 14th of April the Court at Altona tried Mr. Laers, the Finnish minister, for breaking open the room and trunk of Jacob Jongh, who had eloped with his wife. He was also accused of making an inventory of Jongh's property, which ought to have been done by the vice-director and the court, and that by so doing it was alleged he "usurped," "suspended" and "vilified their authority." He was sentenced to make good to the Company what Jongh owed, viz.: 200 guilders in coin, 40 guilders in beaver; besides a fine of 40 guilders for having "vilified authority." He was also tried for having married himself, which the Court said "was directly contrary to the orders sanctioned about marriage connections." His marriage was therefore declared "null, void and illegal." Laers sent in a statement to the Court, in which he says he had been condemned to "heavy amends," which, in his "poor situation," he could by no means pay; that he "paid already nearly 200 guilders," and now was again condemned to pay a fine of 280 guilders. He also alleged that he only broke open the door because he was informed his wife was concealing herself there, and found nothing but a few pairs of stockings, which the fugitive raptor of his wife had left behind." He said it was his "submissive petition" that Beekman "would make a favorable and merciful intercession in his behalf," and pardon graciously what was committed "through ignorance," and to save his "reputation and condition as a minister not to inflict any punishment." With regard to his marrying himself he said: "I cannot discover anything illegal in it. I acted just in the same manner as I had done before in respect to others; exactly as others do who are not prosecuted for it, and I can conscientiously assure you that it was done without any evil intention. Had I known that my marrying myself in this manner should have been so unfavorably interpreted, I should have submitted to the usage of the Reformed Church. But I did not know it. Wherefore I pray once more the honorable general that he will vouchsafe me his aid, and take into consideration my forlorn situation, so that I, without becoming a burden to others, may supply my daily wants." As to whether the Rev. Mr. Laers was pardoned, or had his fine remitted, history is silent.

But it is not silent in regard to the injustice and oppression of Beekman. He has condemned himself by his own writing. According to the 7th article of the terms of the surrender of Fort Christina, the Swedes and Finns remaining on the South river were to be allowed the liberty of their national religion, known as the Augsburg Confession. Laers was not compelled to conform to the usages of the Reformed Church in marriage; but to his own. The breaking open of a room and trunk to see if his wife was in the one and his own goods (of which he was robbed by an adulterer) in the other were acts liable to be committed by any man under such circumstances, and the assumption that Jongh had left behind him in that trunk goods enough to pay the debts of the Company, and that they had been taken and kept by Laers, was most unwarrantable. It should have occurred to the meanest intellect that if the adulterer had taken with him the goods of the husband whom he had wronged, he would also hardly have failed to have carried off those belonging to himself when they were portable enough to be contained in a trunk. It was one of the rankest cases of judicial injustice ever committed on the Delaware.

At this time we see mention made of the erection of the first mill in this State north of the Christiana. It was situated on Turtle, or Skilpadle, creek (now called the Shelpot), about one and a half Dutch, or four and a half English, miles from Altona. John Staelcup petitioned for the land around the mill, so that he "could attend to it with greater safety."

This mill stood in 1769, and was then carried on by a Swede, or Holsteiner, named *Tapoeise.* Dankers and Sluyter, the Labadists, visited it that year whilst passing through Delaware to visit Augustin Herman in Maryland. They describe Tapoeise as short in person, but "a very friendly fellow," having several children. He is the first miller in our State of whom history records a description.[1]

A grant of land at the Hoernkills was made to a party of Mennonists. The association was to consist

[1] See Travels of Dankers and Sluyter, in Memoirs of Long Island Historical Society. (The Labadists were a sect of Christians that lived at one time in Delaware and Maryland.) Also, O'Callighan, vol. 2, p. 466.

of married males and single men who had attained the age of twenty-four years, who were not bound to service or indebted to the association. No superiority or office was to be sought for; but all persons were to obey the ordinances for the "maintainance of peace and concord." No minister of the gospel was to be allowed in the association, for being composed of persons of various religious opinions, no one minister could preach in accordance with the sentiments of the whole of their community, and to get one of each sect, it was argued, would not only be impossible, "but an inevitable pest to all peace and union." The number that agreed to settle was thirty-five men, the principal of whom was Pieter Cornelioz Plockhoy, of Zierikzee. The City of Amsterdam agreed to loan each of them one hundred guilders. The whole community were to be security for this loan. Thus every man was surety for all the rest. It is not made manifest with any degree of certainty on the records, whether or not any of this association ever emigrated to Delaware.[1]

In the early part of September, New Amstel was again alarmed by fears of an attack from the savages. An Indian came running into the town severely wounded, to the consternation of the inhabitants. He gave them to understand, as far as he could, that it was done by the Senecas. A short time afterwards an old man, named Jans Flons, whilst riding in the woods with his wagon and two horses, was shot and

[1] O'Callaghan, vol. 2, p. 466. Broadhead and O'Callaghan, vol. 2, pp. 176, 177.

killed. The inhabitants immediately fled with their property to the fort. They apprehended an attack; but they were not further molested.

On the 17th of November a youth, the servant of John Stalcup, was murdered about four hundred rods from Fort Altona. His master had just left him. These murders, it was supposed, were committed by the Senecas. A small house was at another time burnt by the savages, near New Amstel.[1]

On the 3d of December three Minqua chiefs visited Altona. They informed Beekman that the murderer of the youth was a young Seneca captive. They also informed him that as long as the Christians had resided on the river, they had never in any manner injured or offended them (the Minquas); on the contrary, they showed them every mark of friendship, and were always willingly and cheerfully employed in reconciling differences between them and the other savages. They also told Beekman that about three years ago one of their nation was murdered by Christians at New Amstel; but they did not resent it. That they expected ere long to their assistance about eight hundred Swedish Minquas, of whom about two hundred had arrived, so that the next spring they were resolved to go and make war on the Senecas and attack them in their forts. They solicited the Dutch to supply them with ammunition when they paid for it.

On the 29th of March D'Hinoyossa granted the sole privilege of trading between Bombay Hook and

[1] Albany Records, vol. 17, pp. 230, 245, 246.

Cape Henlopen to Peter Alricks. The penalty for violating this privilege was forfeiture of goods. This order caused much discontent amongst the Swedes. He also prohibited the selling of liquor to the Indians under a penalty of 300 guilders. If any one sold the Indians liquor, they (the Indians) were to be allowed to rob them of it.[1]

The disputes between Beekman and D'Hinoyossa still continued. Beekman, in his letters to Stuyvesant, constantly reiterated his complaints of the conduct of D'Hinoyossa, who insisted on vessels lowering their colors when they passed New Amstel, and threatened to examine their cargo. Beekman, claiming the jurisdiction over the river, denied his right to do this, and in May summoned him to appear before him at Altona. This summons D'Hinoyossa disregarded. Beekman further charged him with making, in a tavern, attacks on Stuyvesant. Also with charging the inhabitants of New Amsterdam with bringing the Swedes to South river by their ill treatment of Minuit, and threatening, if he could, to drown or poison the Manhattans.[1]

In a letter to Stuyvesant in June, he charges D'Hinoyossa with taking away the palisades of the fort and burning them in his brewery; also, with selling to the savages the new city guns which arrived in the ship Parmeland Church; also, with selling to the English in Maryland the city millstones, brought in the same ship, for one thousand pounds of tobacco, and a small brewer's kettle for seven or eight hundred

[1] Albany Records, O'Call., vol. 2, p. 465.

pounds; also, with railing against the Manhattans and threatening vengeance.[1] These charges Beekman sustained by affidavits.

In July Stuyvesant proclaimed a day of fasting and prayer in New Netherlands, and Beekman complained that D'Hinoyossa did not publish it in his jurisdiction, giving as a reason that his colony was not "especially mentioned." About the same time D'Hinoyossa also issued a proclamation for fasting and prayer, but in the name of the Director and Council of New Amstel (thus ignoring the power of Stuyvesant); but which, said Beekman, they "*did not keep.*" D'Hinoyossa also wrote to Beekman, informing him that he expected from them (Stuyvesant and the other officers of the company) "no more similar orders or injunctions;" but that "in future" they "would take care of it" for themselves, and for that purpose they had "established a quarterly prayer day."[2]

In a letter written by Beekman in August he says "he cannot live on good terms with D'Hinoyossa." Stuyvesant also severely censured his conduct in a letter to the directors in Holland. They had written to Stuyvesant, complaining "of certain proposed proposals, and defence against the savages, and a divisional line by Maryland," which they said they never passed. Stuyvesant, in a letter dated July 15, in reply, said: "As to what passed between Maryland and D'Hinoyossa remains a secret. Beekman gave, it is true, some communications that in conse-

[1] Albany Records, vol. 17, pp. 158, 159. [2] Ibid. p. 195.

quence of the murder of two or three English, going from New Amstel to Maryland, trouble and some war had arisen between them, which caused certain correspondence, and some embassies sent *vice versa* by D'Hinoyossa and the Governor of Maryland; but all this without any action with us or Beekman, so we are ignorant about it. *Daily occurrences prove that D'Hinoyossa is either too ignorant or too great in his own eyes to consult your ministers on those or other such like affairs.*"

In September there was a runaway servant hung, the first execution we have any record of in this State. From the loose structure of the language it is almost impossible to say who hung him or for what he was hung. It rather conveys the impression that he was hung by the Indians; but a letter from the directors in Holland to Stuyvesant, written April 16, 1663, gives the impression that D'Hinoyossa hung him for the crime of running away, though this it is difficult to believe.

Beekman, writing to Stuyvesant, says: "On the 19th was hung, the head cut off and placed on a stake, in the presence of "French," one of the English runaway servants, and bought by Peter Alricks at Hoernkill from the savages. When these were to be conducted to New Amstel by some English, French made an attack upon them near Bompjes (Bombay) Hook, wounded two, when they fled out of the boat; but were again overtaken at New Amstel, where they were apprehended by orders of D'Hinoyossa, when on the 3d the English masters departed; but D'Hinoyossa refused to deliver French to them, because he committed, as he suspected, a delict in the distance of the colony. I sustained that the case belonged to your honor's jurisdiction, as the deed was committed on the river and not in the colony. Van Sweringen was sitting as judge in the case."[1] The directors in Holland, writing to Stuyvesant, say: "The complaints which have been made against the director of this city, Alexander D'Hinoyossa, and the sheriff, Gerritt Van Sweringen, and their proceedings there in *executing and hanging a runaway servant who resided before in New Netherlands, is strange.*"[2]

In November D'Hinoyossa and Van Sweringen left suddenly in a shallop to meet Calvert, the Governor of Maryland, at the house of Augustin Herman. They went there in obedience to a request by letter from Calvert. They gave no information of this movement either to Beekman or Stuyvesant. The information in relation to this meeting was sent to Beekman by a Mr. J. Willems, who appeared to be a sort of spy in the service of Beekman at New Amstel. Willems, like most spies, determined to earn his wages. He narrowly watched every action of D'Hinoyossa and reported them all to Beekman. He described a meeting in the valley near Fort New Amstel, where D'Hinoyossa and several persons who were with him "lifted up their eyes towards heaven, laid their hands upon their breasts," and did several other unnecessary and unmeaning things, from which Willems suspected they were plotting evil to the com-

[1] Albany Records, vol. 17, pp. 243, 244. [2] Ibid. 128.

pany of the colony. This Willems died shortly afterwards.

Mrs. Pappegoya, the daughter of Governor Printz, who had since lived at her father's place at Tinnicum, where was situated the Finnish village of Printzdorp, sold it to Mr. LaGrange. A part of the purchase money was a bill of exchange, which was protested. Suit was entered on this bill of exchange before Beekman at Altona, who gave judgment against LaGrange. This was the first known protested bill on the Delaware, and the first case tried on any written obligation within the limits of the territory constituting this State. An appeal was taken, and Tinnicum island and the Printz property was afterwards the cause of a great deal of litigation between the Mrs. Pappegoya and others. There is some evidence to show that at this time her husband, Pappegoya, had died. Mrs. Pappegoya lived alone, and the fact of his decease is mentioned by the Labadists, Dankers and Sluyter, who visited Tinnicum in 1679. But history neither records the time, nor place, nor circumstances of the death of the fourth governor of Delaware.

[1663] The first letter written by Beekman this year was one filled with the usual complaints of the conduct of D'Hinoyossa. "He sells everything for which he can find a purchaser," says Beekman; "even powder and musket balls from the magazine. I know that he sold a considerable quantity to Augustin Herman, besides nails belonging to the city," &c. And again he says: "D'Hinoyossa considers us his mortal enemies." He rejected a surgeon because he was a friend of Beekman.

In the early part of February the small-pox appeared amongst the savages on the South river. This is the first mention of small-pox on the Delaware. We have now, too, mention made of the first Frenchman in Delaware. His name was Pickard. He owned a house and land in Altona, which he sold and then removed away.[1]

In February also the duty of four stivers on a beaver, levied on the trade of South river by the company, was abandoned, and left for the benefit of New Amstel.

The records, whilst they abound with the complaints of Beekman on the conduct of D'Hinoyossa, show little or no complaint of D'Hinoyossa against him. But from this it must not be supposed that his letters showed one mite more of satisfaction at the acts of Beekman, than those of Beekman did of him. Beekman's letters, which so minutely detail the history of the territory that now constitutes this State whilst under the rule of the Dutch, were written to Stuyvesant, in New Amsterdam, and are therefore preserved in the New York Archives; whilst those of D'Hinoyossa were written to the Burgomasters of Amsterdam, in Holland, and were not therefore as well kept, or so conveniently within our reach. It is more than probable they censured the conduct of Beekman as bitterly as those of Beekman did his. Now and then a complaint of D'Hinoyossa was laid

[1] Albany Records, vol. 17, p. 260.

by the Burgomasters of Amsterdam before the Directors of the Company. In a letter from them to Stuyvesant, they allude to "the repeated complaints of the commissaries that the Company's government in New Netherlands seemed to make it a study to oppose the colony of the city—to prevent its growth," and alleged, amongst other examples, that they "did protect the colonists who, from time to time, escaped from the city's colony, and hired by sinister means their best and ablest farmers from there." This, as they alleged, had actually been practised by one Brex Wolters. "Said commissaries entreated us most seriously, as they had again concluded to send there fifty laborers and twelve girls for the service of the colony, not only that similar measures might not again be put in practice, but further, if any of them within three years might escape from the service in the colony to the Manhattans, they might be sent back again."

On the 17th of May a grant of a small valley, situated on the west side of Fort Altona, together with some woodland, in all eight morgans, or fifty six acres, was made to Beekman by Stuyvesant. This is the first recorded grant of land to an individual within the limits of the city of Wilmington that we have an account of. From its description we should judge that it extended from Church street to Walnut street, and from the Christiana probably to near Ninth street, as the grants generally at that time far exceeded the estimated measurement, and a great part of what is now firm land was covered by the Christiana, as well at low as at high tide. The hill that commences at Walnut street would form a valley between Fort Christiana and French street.[1]

In the early part of June a battle took place between the Minqua and Seneca Indians. "The Senecas, to the number of eight hundred, blockaded the Minquas in their fort whilst a large proportion of their number were out hunting. When the Senecas approached, three or four men were despatched to the fort with the offer of peace, while their force remained at a distance; but a Minqua returning from hunting discovered the Senecas, so that the next day those in the fort concluded to meet them with twenty or thirty men. The other Minquas at the same time, with their forces, made an attack, put the Senecas to flight, and pursued them for two days, retaking ten persons and killing ten Senecas." The Governor of Maryland assisted the Minquas with two cannon and four men to manage them. The accounts of this battle handed down to us are confused. It is more than probable the assistance rendered by the Marylanders contributed to the Minquas victory. The site of the battle is not definitely known; but it is supposed to have been within the limits of the State. Probably in the neighborhood of Iron or Chestnut Hill, near Newark, as the Minquas fort was situated on a high mountain. These hills answer best to the description given by Campanius as the site of the Minquas stronghold.

Gerritt Van Sweringen, the sheriff and counsellor, had shot and killed an insolent soldier. On the 1st

[1] Albany Records, vol. 21, p. 108.

of June a vessel arrived from Holland, pardoning him for the offence.[1]

About this time we are inclined to believe that the Governor of Maryland fitted out an expedition to attack the Dutch settlement at the Hoernkill. It is generally supposed to have taken place in 1661; but it was more than probable this year. As the Marylanders advanced the Dutch settlers withdrew and abandoned the Hoernkill. Duties were also levied on the trade of the South river at the Hoernkill at this time,[2] and encouragement was given by Baltimore to those who desired to make settlements there, and Col. William Stevens was authorized by him to induce emigrants to take up land in the vicinity.[3]

On the 9th of August Lord Baltimore visited New Amstel, with a suit consisting of twenty-six or twenty-seven persons. On the 11th and 12th of August he visited Altona, where he was entertained by Beekman. Stuyvesant made him an offer, through Beekman, of a convoy and horses if he visited Manhattan. He returned his thanks for the offer, and promised if he visited Boston in the spring, a matter he had in contemplation, to go by way of Manhattan.

In August a vessel arrived at New Amstel from New Amsterdam with farmer's instruments of industry and ammunition for the City's Colony.[4] This was probably the ship Jacob.[5] The same month Beekman

[1] Albany Records, vol. 17, p. 280.

[2] Griffith's Sketch of the History of Maryland, p. 22.

[3] Neill's Terræ Mariæ, p. 161.

[4] Smith's New York, p. 113.

[5] Albany Records,

informed Stuyvesant that Fort Altona (Christina) was out of repair, that the palisades and the whole fort was in decay.

The difficulties between the officers of the Company at Altona and the Colony of the City at New Amstel induced the City of Amsterdam to endeavor to get Altona assigned to them, and have the whole of the South river under their jurisdiction. They had made a proposition to the Director of the Company to that effect in February, in which, amongst other articles, they stipulated that when Altona should be assigned to them, their jurisdiction and property of the land should be from where the South river entered the sea to its head. Their property on the west side of the South river they desired should extend to Baltimore's line. On the east, or New Jersey side, they wished it extended from the river three miles into the country; and that to this territory the Company "should renounce and quit every claim;" that they (the City of Amsterdam) should have "all the rights of high and low jurisdiction which the Company possessed," *provided* that they paid to the Company "the recognitions which were actually paid from all imports and exports, without being obliged to bring their goods into the Company's magazine." They desired "not to be visited by the Company's commissaries;" but "that the Company should remain satisfied with the declarations" of the City's "commissary or director," and that "nothing should be paid to the South river;" neither should "the Company claim any authority upon it." The Company at first declined to grant the whole of the colony, although they offered to assign to the city Fort Christina. They also refused to agree to the non-bringing of the goods into the Company's magazines, and the non-visiting of the city's cargoes by the Company's officers, and the paying of nothing on the South river; also to some others of the articles.

The dangers from the English, however, both north and south, finally induced them to cede the balance of the South river to the City of Amsterdam. Accordingly, on the 11th of September the agreement between them was entered into, and on the 22d of December Stuyvesant made a formal transfer of Altona to D'Hinoyossa, as the agent of that city. D'Hinoyossa had left Holland some months previously, and on the 3d of September had returned to the South river, in company with Peter Alricks, with one hundred and fifty passengers, nearly all of whom were Swedes and Finns. There were thirty-two of the latter nation.[1]

In making this transfer of the remaining portion of the State of Delaware to the City of Amsterdam, the West India Company thought that that city would "prove a wall between them and the English of Maryland;" would increase the influence of New Netherlands with the States General and prevent the bickerings between the officers of the city and Company in relation to jurisdiction, as the city would thus have sole control of the South river, by which the remaining portions of our State, as well as other

[1] Broadhead and O'Callighan, vol. 2, p. 233.

territory, was transferred from the possession of the West India Company to the City of Amsterdam. It comprised all that portion of Delaware north of the Christina, and consisted of the City of Wilmington and the Hundreds of Brandywine, Christina, Mill Creek, White Clay Creek, and a small part of Pencader.

The territory that constituted the State thus passed entirely under the control of the City of Amsterdam, with the exception of that portion that lies between Cape Henlopen and the Maryland line that fronts on the Atlantic ocean, and which was more than probable at this time under the jurisdiction of Baltimore. The following is a copy of the deed of grant:

"Peter Stuyvesant, in behalf of their High, Mighty Lords, State General of the United Netherlands, and the Lords Directors of the Council, attest and declare how we this day, in conformity with their orders and letters of aforesaid Lords Directors of the 11th of September, 1663, transported, surrendered and transferred to the Honorable Alexander D'Hinoyossa, in behalf of the noble, great and respected Lords, Burgomasters of the City of Amsterdam, Director in their colony on the South river, from the sea upwards, so far as that river extends itself, towards the country on the east side, three miles from the borders of the river, and towards the west side so far as the country is extending till it reaches the English Colonies, with all the streams, hills, creeks, harbors, bays, and parts relative to it, of all which land, with all its dependencies, especially so of the Fortress of Altona, we, in the name and behalf of the aforesaid Lords Directors, make at this time a full cession and transfer to the aforesaid Alexander D'Hinoyossa, on behalf of the noble, great, and respectable Burgomasters of the City of Amsterdam aforesaid, making to him a cession of all actual and real possession, property and privileges; and all this upon such conditions as have been agreed upon between the noble Lords Directors and others and the noble, great and respectful Burgomasters of the City of Amsterdam, without reserving any actual or real pretensions, promising therefore to consider and keep this transfer inviolate in truth; wherefore we signed it and confirmed it with our seal in red wax, imprinted in the Fort Amsterdam, in New Netherlands, 22d December, 1663."[1]

The Company had agreed, before they resolved to transfer New Amstel, not to admit within their jurisdiction "any colonists of the city, or its laborers," except they "could show their pass and prove by certificate that the city had received payment of her advance money." They instructed Stuyvesant to consider private debts as "personal matters." The City, in its agreement with the Company, agreed to settle on the South river *four hundred* colonists and other useful husbandmen, and held out the impression that they would send out a greater number. This, said the Directors, in a communication to Stuyvesant, "must contribute to our security against the English north." "Besides," said they, "we may expect a more powerful intercession of the city by our government to obtain from the Crown of England the final

[1] Albany Records, vol. 21, p. 445.

settlement of the long desired boundaries." In obtaining the assignment of Altona, the City of Amsterdam "reserved to herself alone" the "exclusive commerce" of the South river. This alarmed several of the merchants of New Amsterdam, for fear the commerce of the Manhattans might be diverted to another channel. This fear, however, was without foundation, as the intention of the City of Amsterdam, by reserving the exclusive commerce of this river, "was only to insure cargoes for the ships which she sent over with the laborers and colonists."

The West India Company could hardly have been aware of the extent of their grant to the city. It extended up the Delaware to near the Catskill Mountains in New York, in latitude between 41 and 42 degrees; and as Baltimore's grant did not extend beyond the 40th degree of north latitude, the City of Amsterdam would by this deed have obtained possession of nearly the whole of Pennsylvania and a large portion of the State of New York, and might even have extended to the Pacific unless the Virginians had interfered with them in the progress westward.

On the South river at this time, according to the report of the Commissioners of the City of Amsterdam, the Swedes, Finns, and other nations had established about 110 good boweries or farms, which had a stock of 2,000 cows and oxen, 20 horses, 80 sheep, and several thousand swine. It was recommended that no Hollander should be employed in agriculture; but that Swedes, Finns, and other foreign nations should be induced to emigrate to the South river for that purpose. The city was to offer to lend such people sufficient to pay their passage and purchase agricultural implements.[1]

Most of the emigrants who arrived in the Parmeland Church with D'Hinoyossa were Swedes and Finns, who were aided by the City of Amsterdam in this manner.

The Dutch of Delaware at this time brewed a great deal of strong beer, which was sold to the Marylanders (who did not manufacture any) for tobacco.

On the 4th of November Andreas Hudde, who figured so prominently in the early portion of our history, died at Appoquinimy, which was then the name of Appoquiniminck. He had been a faithful servant of the Dutch for many years, and his services were appreciated by them; but he had been robbed and all his property destroyed by the Indians, and he had sunk from the position of commissary, or governor, to that of clerk. He petitioned for his discharge as clerk, and it being granted, had left Altona on the 1st of November and was going by way of Appoquinimy to Maryland, where he intended engaging in the brewing business; but he died before he reached there of an "ardent fever." His first service under the Dutch was as surveyor at Manhattan, 1642, from which station he was removed; in 1645 he was commissary of Fort Nassau, since which time he had been identified with the Dutch interest on the South river.[2]

[1] Broadhead and O'Callighan, vol. 2, pp. 210, 211.
[2] Albany Records, vol. 17, p. 309.

CHAPTER XXVI.

1664.

Resignation of Beekman—Absolves Swedes from their oaths—They are summoned to New Amstel to take a new oath—They refuse—Allowed eight days to leave or take the oath—Farm offered to Beekman—He declines—He begs office of Stuyvesant—Is made sheriff of Esopus—D'Hinoyossa sole governor of South river—Intends building his capitol at Appoquiniminy (supposed to be Odessa)—Intends constructing dykes around valleys—Appoints Alricks superintendent of fur trade—Prohibits brewing and distilling—All goods exported to pay recognitions—Order of boundaries—Swedes fit out an expedition to recover Delaware—Vessels run ashore—Expedition abandoned—Grant of territory between the Delaware and St. Croix to Duke of York—Under it no grant of Delaware—Duke of York grants New Jersey to Berkley and Carteret—Swedes present memorial to States of Holland in relation to New Sweden—Expedition fitted out under Col. Nichols—Slaves for New Amstel arrive in New Amsterdam—Capture of Manhattan by Col. Nichols—Name of New Amsterdam changed to New York—Narrow escape of New Amstel slaves—Surrender of town of New Amstel to Sir Robert Carr—Terms—D'Hinoyossa refuses to surrender—Retires to the fort—Storming of the fort—Three of the garrison killed and ten wounded—It is plundered—Name of New Amstel changed to New Castle—The South river to Delaware—Fort Altona to its old name of Christina—Destruction of Fort New Amstel (Casimer)—Slaves divided amongst captors—Murders by Indians—Momentous consequences of fall of Dutch power in Delaware—Author's prediction—Destiny of Anglo Saxon race—Plan of union of (note) Confiscation of property of D'Hinoyossa, Alricks and Van Sweringen—D'Hinoyossa and Van Sweringen retire to Maryland—D'Hinoyossa's letter—Goes to Holland, &c.

[1664] Although Altona had been transferred to the City of Amsterdam, and it and New

Amstel were both under the governorship of Alexander D'Hinoyossa, Beekman continued to act as Governor of Altona until the 5th of June, when the Swedes and Finns were summoned to appear at Fort Altona, and he then resigned his office.

On the 9th of June the Swedish deputies and a great part of the Finns assembled at Altona, and at their request were absolved by Beekman from the previous oath they had taken. They were much dissatisfied at the change and made use of the expression: "We are now sold—deliver us." A new oath was proposed to them by the directors and council on the 8th of June, in the presence of Beekman, to be taken at New Amstel. This they unanimously refused to take, unless they received previously in writing a promise of "*all such privileges of free trade and others as they were favored with under the Company's administration.*" Without this they said they would be "compelled to depart." Eight days were allowed them for consideration, when all who did not take the oath would be compelled to leave the South river.[1]

D'Hinoyossa, under the City of Amsterdam, was now the sole governor of the whole South river, on both its eastern and western shores, and Beekman was out of office. He, however, offered to Beekman, if he would stay at Altona and take possession of the great farm in the vicinity of the fort, to provide him with some five or six men servants. This Beekman declined, as he could see no good that would result

[1] Albany Records.

from it, "as in future," he said, "no freeman would be permitted to trade either with the English or savages, and the tobacco or fur trade was reserved for the city."

He wrote pitifully to Stuyvesant for an office, telling him he "could not live" on the South river "as a freeman and support his family, and did not wish to go to Maryland." He wound up his appeal by saying, "take care of me, father."[1]

A short time after this Beekman took his departure for Esopus, of which place he was appointed sheriff. He was continued in office under the English government, and afterwards became an alderman in the city of New York, where one of the principal streets (Beekman street) is named after him. He died in that city, at the age of 85, in the year 1707. He must have had a numerous family, as during his brief residence at Altona he had three sons born. A daughter of his married a son of Stuyvesant. In speaking of him, Acrelius says:

"While the colonies were kept up, Beekman had a share in the administration; but was little considered. This made him envious of D'Hinoyossa, whom he aspersed in frequent letters to Stuyvesant."

Many of his charges are, however, supported by evidence. His letters to Stuyvesant form a most minute history of affairs in this State at that time. After his removal to Esopus they, of course, ceased.

D'Hinoyossa, now the sole governor on the Dela-

[1] Letter from Wm. Beekman to Peter Stuyvesant. See Hazard's Annals, p. 355.

ware, determined to take up his residence in Appoquinimy, and there build his capital. Here it was determined to erect a large stone fort and promote trade with the English.[1] He resolved to construct dykes around several valleys in that vicinity, which Beekman, before he left, informed Stuyvesant "the savages would be much opposed to." He appointed Peter Alricks to superintend the fur trade and reside at New Amstel; a man named Israel, a member of the Council, to reside at the upper end of Passajong (changed to Passyunk—now part of the 1st and 26th Wards in the city of Philadelphia), and another member of the Council to superintend the trade at the Hoernkill. He also prohibited brewing and distilling in the colony even for domestic use, and ordered that all goods and tobacco exported should pay recognition.[2]

The slave trade was now engaged in. The city of Amsterdam entered into partnership with the West India Company to engage in this traffic. On the 20th of January they made a contract with Lymen Gylde to bring from Loango, on the coast of Africa, in the ship Gideon, three hundred slaves. The city was to receive one-fourth of these slaves when they arrived for the use of the colony on the South river.[3]

On the 23d of January an order issued by the State General claimed that New Netherlands on the South river extended sixteen leagues south of Cape

[1] D'Hinoyossa's capital was undoubtedly the present village of Odessa, at the end of Herman's Cart Road.

[2] Albany Records, vol. 17, pp. 317, 318.

[3] Broadhead and O'Callighan, vol. 2, p. 222.

Henlopen. The Swedes, who had apparently abandoned their possessions on the South river, now fitted out an expedition for their recovery. The Dutch, however, who seem to have had efficient spies in Sweden, were early informed of their intention, and in a letter dated the 16th of October, 1663, had given information of the matter to Stuyvesant.

This expedition consisted of a frigate called the Falcon, of thirty-two guns, and a yacht of eight or ten guns, manned, in addition to the customary crew, by two hundred or more soldiers, who were to be taken on whilst passing the Sound. This expedition was commanded by the Swedish Vice-Admiral, Hendrick Gerritson Lesselin, or Lechelm, who was well acquainted with New Netherlands, having been employed there in 1641 in the ship Neptunis, from Curacao.

Had it arrived at the South river, the Dutch would have been able to have made but a poor defense, as both Forts Altona and New Amstel were in a state of decay, having but few pieces of cannon. The Swedes and Finns, too, on the Delaware, and who were dissatisfied with their rule, far outnumbered the Dutch inhabitants. Indeed, the latest emigrants, who arrived on the ship Parmeland Church, were nearly all Swedes and Finns, who could not be expected to fight against their own countrymen. This expedition, whilst on its way to New Netherlands, met with such disasters that it had to be abandoned. It first ran aground on the Landts Croon, situated near Heisselberg, in Denmark, at the north entrance of the Sound, and having miraculously got off and passed the Sound, it again ran aground on what is supposed to be the Island of Anholt, where the smallest vessel was wrecked, with all her stores. The Falcon unshipped her compasses and found it necessary to run for Gottenberg, when she again went ashore and was so injured that she had to be dismantled. Her crew were, therefore, discharged, and the expedition given up. Thus ended the last attempt of the Swedes by force of arms to recover their possessions on the South river.

Stuyvesant did not think the City and Company exerted themselves sufficiently to resist this threatened attack. In writing to them in relation to this expedition, he says: "It excites in ourselves strange emotion that your Honors and the Worshipful Commissioners over the City's Colony, who have had such long and particular knowledge and information of this meditated expedition, and did not, conjointly, immediately and instantly request and apply to the Lords of the Admiralty residing at Amsterdam for a man-of-war sufficiently powerful and fit to counteract so ruinous a design."[1]

The Swedes, on the failure of their expedition, determined to see what could be done by diplomacy. Accordingly the Swedish Ambassador, Harald Appelboom, proceeded to address a series of energetic notes to the States General, demanding the restitution of New Sweden. In one of the 19th of June, on this

[1] Broadhead and O'Callighan, vol. 2, pp. 232, 236, and Albany Records, vol. 4, p. 466.

subject, he requested "prompt expedition, reparation and satisfaction in the complaint so frequently made, and again repeated." He again, on June 27th, makes another demand, in which he recites the Swedish claim. In this he alleges the right of Sweden to the western side of the Delaware. He acknowledges the right of the Dutch to the eastern side. In this note he complains of the Dutch "debauching not only the Swedish inhabitants who happened to be" on the South river; "but even drawing and conveying from Finnland and Old Sweden additional inhabitants to be employed in their service in New Sweden." He states in one of these notes that "the Swedish people are more conversant with and understand better than any other nation the cultivation of pasture, wood and tillage land, fishing, hunting and fowling."

This last complaint of Appelboom's must have alluded to the Swedes and Finns who sailed with D'Hinoyossa in the Parmerland Church.[1]

Upon the receipt of these repeated remonstrances and demands for redress from the Swedish Government, the States General referred the matter to the West India Company for information, and they on the 9th of October made answer through their secretary, Michael Tenhove, that they had the best right to the South river. In this answer they recited their original discovery and possession, and the capture of Fort Cassimer and other outrages done to them by the Swedes. They also stated that having sold the territory in dispute to the city of Amsterdam, they had,

[1] Broadhead and O'Callighan, vol. 2, pp. 239, 240, 242.

therefore, no longer possession of the place.[1] This was the last demand of the Swedes for the possession of their American territories that history makes mention of; for now occurred an event which was to render nugatory and futile all attempts of the Swedes, whether by arms or diplomacy, for the possession of New Sweden, and also all the enterprises of D'Hinoyossa for the benefit of the Dutch trade on the South river. This was the happening of what Stuyvesant had long feared and predicted, viz., the conquest of New Netherlands by the English. D'Hinoyossa had scarcely enjoyed his power four months when it was wrested from him and he himself was expelled from New Netherlands.

On the 12th of March[2] Charles, II., king of England, granted to his brother James, Duke of York and Albany, a patent for all the land embraced between the river St. Croix (which is now the northern boundary of the United States) and the east side of the Delaware bay. This grant really comprised what is now the New England States and the States of New York and New Jersey. It took in all of New Netherlands belonging to the West India Company and the possessions of the City of Amsterdam on the east side of the Delaware; but it did not take in the City's Colony, or what was then known as New Sweden, which was situated on the west side of the Delaware. Nearly the whole of this, or from a point a little north of New York, or Patterson, New Jersey, to the

[1] Broadhead and O'Callighan, vol. 2, p. 259.

[2] English dates are old style.

river St. Croix, had been granted in 1589 to the Plymouth or North Virginia Company by his grandfather, James I.,[1] whilst the west side of the Delaware, from Philadelphia to the Virginia line, or from the 38th to the 40th degree of north latitude on the peninsula between the Chesapeake bay, the Susquehanna river and the Delaware, had been granted by his father, Charles. I., to Cecilius, Lord Baltimore. Yet, under this grant, whose metes and bounds are so distinctly defined, James, Duke of York, conquered and held possession of the territory now comprised in the State of Delaware, which was not mentioned in his deed, but which was distinctly mentioned and set forth in the grant of his father, Charles I., to Baltimore.

On the 3d and 4th of July[2] the Duke of York granted the territory between the Hudson and Delaware rivers to Lord Berkley and Sir George Carteret. The territory comprised in this grant now constitutes the State of New Jersey.

When the grant of this territory was made to the Duke of York, war had not been declared by the English against the Dutch; but it was soon afterwards. On the 25th day of May (old style) an expedition sailed from Portsmouth, England, for the purpose of capturing all the Dutch possessions on the continent of North America known as New Netherlands. It consisted of four vessels, viz., the Guinea, a frigate of thirty-six guns, commanded by Captain

[1] See pages 90, 91, ante.

[2] The reader must bear in mind that English dates are old style.

Hyde; the Elias, of forty-two guns; a ship of eighteen guns, and a transport of fourteen or sixteen guns. On these vessels, in addition to their crews, were three hundred soldiers. The expedition was commanded by Colonel Richard Nichols, and with him came Sir Robert Carr, Knight; George Cartwright and Samuel Maverick, Esqs., with extensive powers to visit the several English colonies and to hear and receive, examine and determine all complaints and appeals in all matters, military, civil and criminal, and to provide in all things for "settling the peace and security of the country." They were also instructed that the first business to be done was the "reducing of the Dutch in or near Long Island, or anywhere within the limits of" the English "dominions, to an entire obedience." Their instructions recited that the Dutch colonies "were a receptacle and sanctuary for mutinous, seditious and discontented persons, who fly from justice as malefactors, and who run away from their masters, or avoid paying their debts, or who have other wicked designs." They also say that as soon as they (the Dutch) "grow to any strength or power," "their business is to oppress their neighbors and engross the whole trade to themselves" by any indirect, unlawful, or foul means. In proof of this the instructions recited "their inhuman proceedings at Amboyna,[1] in a time of full peace and all professions of particular love and friendship." Reciting this, the instructions

[1] The English were cruelly massacred and tortured there by the Dutch, for which Cromwell afterwards compelled them to pay about $300,000.

say "it was high time to put them (the Dutch) without a capacity of doing the same mischief" in America, "and reducing them to the same rule and obedience" with the English "subjects there."

The Commissioners were instructed to use no more violence than was necessary to reduce them to obedience. No man who "would yield obedience" was to be "disturbed or removed from what he possessed." Those who would submit to English authority were to enjoy the same privileges as English subjects. They also said that they (the Dutch) had "no kind of right to hold what they were in possession of," as "they were King Charles' unquestionable territories, which they were possessed of by an invasion of English right."

The Dutch had received some information of the intentions of the English, yet they had made no adequate preparations to counteract their designs. The English of Connecticut were pressing on the Dutch at Long Island. Several English villages were established there, and in December a Captain John Scott had landed on that island, from Connecticut, with a troop of horse and foot, and took possession of part of it, and in some settlements displaced the Dutch magistrates and appointed English in their place. He and his men committed several outrages, amongst which was the beating of a son of Captain Martin Krygier (named after his father, Martin Krygier,[1]) and several others.

Stuyvesant appointed some commissioners to con-

[1] Broadhead and O'Callighan, vol. 2, p. 483.

fer with him, amongst whom was Cornelius Van Ruyven. On the 15th of January, during a conference, he informed Van Ruyven that the Duke of York not only intended to reduce Long Island, but the whole of New Netherlands to obedience, and that he designed fitting out vessels for that purpose.[1]

Information had also been received by the West India Company of the sailing and destination of the expedition in June, and they immediately informed the Council of the City of Amsterdam and urged that city to use its influence with the States General to get them to send a force to aid in the defense of New Netherlands. This the City of Amsterdam resolved to do. However, as the English expedition was well on its way across the Atlantic, as will be seen in the sequel, New Netherlands fell before it would have been possible to have rendered it any effectual assistance, even had the Dutch moved energetically in the matter.

On the 25th of August the frigate Guinea, the first ship of the expedition, arrived in the outer bay of New Amsterdam, information of which was immediately sent to Stuyvesant. The English at once issued a proclamation, offering protection to those who would submit. On the 27th they captured a sloop engaged in the service of Peter Alricks, the Commissary of the South river, who had come from New Amstel, with twelve soldiers, for the purpose of purchasing cattle, which he had succeeded in doing in New England and having them brought to Manhattan, previous to

[1] Broadhead and O'Calligan, vol. 2, pp. 400, 587.

shipping them to New Amstel. Both the cattle on the vessel and those not yet shipped on the shore were captured by the English.

A few days previous to this the ship Gideon had arrived at New Amsterdam from Africa with 290 slaves of both sexes on board, one-fourth of which belonged to New Amstel. New Amstel's portion of them were hastily run in gangs through New Jersey overland to the South river by Alricks, and narrowly escaped capture by the English. The boat in which they crossed the North river, and some of its crew, was taken.

This was the first introduction of slaves into Delaware from Africa of which we have any record. Slaves were on the South river from its earliest settlement; but we have no account of the mode and manner by which they were brought to its shores.

On the 28th the three other vessels composing the English expedition arrived in the bay of New Amsterdam. Stuyvesant in the meantime had placed the fort in the best posture of defence he could and sent to the various villages for assistance. The inhabitants, however, refused to come to his aid, giving as a reason "that they would not leave their wives and children a prey whilst aiding to defend another place."

He also sent to the commanders of the expedition, by the hands of four commissioners, amongst whom were Cornelius Van Ruyven and the Rev. Mr. Megapolinisis, inquiring the object of their arrival and continuance in the harbor without giving notice of their design, which they ought to have done.

The next day (the 30th) his commissioners reported to him that Colonel Nichols demanded the Fort and place; that he said he would not argue the case in relation to the English or Dutch title; but that he would attack Amsterdam in Holland if so ordered. He demanded a surrender of the place without debate. Several days in the meantime elapsed, and Stuyvesant on the 2d of September sent to the English a lengthy letter, in which the Dutch and English titles were elaborately gone into.

On the 4th of September the English came nearer the Fort, when they were joined by Captain Scott, with his horse and foot, numbering about sixty men. Another demand was made to surrender, and two of the English vessels sailed above the Fort. At this Stuyvesant went up to his gunners as if to order them to fire, but the two Megapolinisises (father and son) took hold of him and led him away.

The English had made a peremptory demand for surrender, and Stuyvesant had replied that he had determined to defend the Fort; but the commissioners requested further time for consideration and cessation of hostilities, and Captain Hyde of the Guinea agreed not to fire on the town.

On the 5th a meeting of the inhabitants was held and a remonstrance voted on and passed, which was handed to Stuyvesant signed by most of the citizens, imploring him as there was "no hope of relief," and as it was "impossible to make head against so powerful an enemy, to surrender," or else they "would call down on his head the vengeance of Heaven" for "all the innocent blood which should be shed" on account of his "obstinacy."

The English commander having refused to give another day, and thus worked upon from all sides, Stuyvesant agreed to treat for a surrender. On the 6th the terms were arranged and concluded; on the 8th they were ratified and exchanged; and on the 9th day of September New Amsterdam, its fort, and the whole of Manhattan were formally surrendered to the English. The Dutch marched out with all their arms, colors flying and drums beating, and the English marched in and took possession of the Fort.

The Dutch soldiers were extremely anxious to fight, but were hurried off and placed on board the slave-ship Gideon before the arrival of the English. They demanded powder, and Captain Krygier promised to give it to them; but instead, carried it to his own house, as it was feared they might attack the English if they had it. The burghers were suspicious of the soldiers. They had threatened to plunder the place, and exclamations were heard amongst them such as, "they knew where the booty was to be got, and where the young women were who wore gold chains," and that they now had "an opportunity of peppering the devilish Chinese who had made them smart so much."[1] A crowd of them surrounded the house of Peter Meyer and attempted to plunder it, but were prevented by the burghers.

The terms of capitulation, amongst other articles, provided: That the people should be free citizens and enjoy their lands and goods; those who desired should

[1] Chinese, a nickname applied to petty traders.

be allowed to move away; any one could come from the Netherlands and plant in the country; ships should be allowed to go to the Netherlands and come to Manhattan for the space of six months; the inhabitants to be allowed to traffic with the English as with Indians; inferior magistrates to remain in office; the towns of Manhattan to choose deputies and to have a free voice in public affairs, and no Dutchman or Dutch ship to be pressed to serve in war against any nation; every Dutch soldier staying in the country was to have fifty acres of land.

Fort Amsterdam at the time of its surrender was totally untenable. It was built only as a defense against Indians, and was never intended to stand a siege against a civilized force. Not more than 300 men could be raised capable of bearing arms in Manhattan; there was not 600 pounds (or a day's supply) of powder in the fort; there was little or no provisions; the houses were built close up to the fort, and the bombardment necessary to reduce it would have destroyed New Amsterdam and ruined its citizens, who expected to be pillaged, in the result of resistance, both by the English and their own soldiers.

Again, the citizens were dissatisfied with the West India Company. On the 10th of November both they and Stuyvesant had informed the Company they had expected an attack both from the English and the Indians, and the Company had not deemed their complaints worthy of an answer. Therefore the citizens, when discussing amongst themselves the question of the surrender, publicly exclaimed: "If the honorable

Company give themselves so little concern about the safety of the country and its inhabitants as not to be willing to send a ship of war to its succor in such pressing necessity, or even a letter of advice as to what we may at present depend on and what relief we have to expect, we are powerless, and therefore will not defend the city, to imperil our lives, property, wives and children, without hope of any reinforcement or relief, and to lose all after two or three days resistance." New Amsterdam at this time contained a population of about 1,500 people.[1]

The name of New Amsterdam was changed by the English to that of New York, by which name it is still known. It is now the largest city not only in the United States, but on the American continent, and will undoubtedly, in course of time, be the largest city in the world.

After arranging affairs in New York, the English turned their attention to the South river, and a commission was issued by the other three commissioners who came with the expedition, to Sir Robert Carr, as follows:—

Whereas, we are informed that the Dutch have seated themselves at Delaware bay, on His Majesty of Great Britain's territories, without his knowledge or consent, and that they have fortified themselves there and drawn a great deal of trade thither; and being assured that if they be permitted to go on the

[1] For the particulars of the surrender and the before mentioned circumstances, see Broadhead and O'Callaghan, vol. 2, pp. 367, 369, 375, 410, 411, 483, 504, &c.

gaining of this place will be of small advantage to His Majesty, we, His Majesty's Commissioners, by virtue of His Majesty's commission and instruction to us given, have advised and determined to bring that place, and all strangers thereabout, in obedience to His Majesty; and by these do order and appoint that His Majesty's frigates, the "Guinea" and the "William and Nicholas," and all the soldiers which are not in the Fort, shall, with what speed they conveniently can, go thither under the command of Sir Robert Carr, to reduce the same, willing and commanding all officers, at sea and land, and all soldiers, to obey the said Sir Robert Carr during this expedition.

Given under our hands and seals at the Fort in New York, upon the Island of Manhattan, 3d day of September, 1664.

(Signed) R. NICHOLLS,
G. CARTWRIGHT,
S. MAVERICK.[1]

Sir Robert Carr was instructed when he came near the Dutch Fort to send his boat on shore, to summon the governor and inhabitants to yield obedience to His Majesty as the rightful sovereign of that tract of land, and let them know "that all the planters shall enjoy their farms, houses, lands, goods and chattels, with the same privileges and on the same terms upon which they do now possess them, only that they change their masters, whether they be the West India

[1] Register of Pennsylvania, vol. 1, p. 36, and New York Records.

Company or the City of Amsterdam. To the Swedes you shall remonstrate their happy return under a monarchial government and His Majesty's good inclination to that nation, and to all men who shall comply with His Majesty's rights and title in Delaware without force of arms."

That all cannon, arms and ammunition belonging to the government shall remain to His Majesty.

Future trading to be regulated by rules of Parliament; right of conscience to be guaranteed; for six months all the present magistrates to continue in office, taking oath of allegiance, and their act to be in His Majesty's name.

If Sir Robert finds he cannot reduce the place by force, nor upon these conditions, he may add such as he finds necessary; but if both fail, he is, by a messenger to the Governor of Maryland, to ask aid. After reducing the place, his first care is "to protect the inhabitants from injuries, as well as violence from the soldiers, which may easily be effected if you settle a course for weekly or daily provisions by agreement with the inhabitants, to be satisfied to them either out of the profits, customs, or rents belonging to their present master, or, in case of necessity, from hence."

The laws for the present to remain as to the administration of right and justice.

He is to declare to Lord Baltimore's son and all the English concerned in Maryland, that this great expense to His Majesty, in ships and soldiers, has been incurred solely for the purpose of reducing foreigners in these parts to His Majesty's obedience; but that being reduced at His Majesty's expense, he is commanded to hold "possession for His Majesty's own behoof and right, and that he is willing to unite with the Governor of Maryland in His Majesty's interest on all occasions; and if my Lord Baltimore doth pretend right thereto by his patent (which is a doubtful case), you are to say that you only keep possession till His Majesty is informed and otherwise satisfied."

"In other things," says the instructions, "I must leave you to your discretion and the best advice you can get upon the place."[1]

In compliance with these orders, Carr sailed from New York with the frigate Guinea, Captain Hugh Hyde, and the ship William and Nicholas, Captain Thomas Morley, and after a long and troublesome passage, prolonged by the ignorance of the pilots and the shoalness of the water, arrived at Fort New Amstel on the last day of September (old style). They passed by Fort New Amstel without notice, the better to satisfy the Swedes, who, notwithstanding the Dutch persuasions to the contrary, were soon the friends of the English.

Carr then had a parley with D'Hinoyossa and the Burghers of New Amsterdam. The burghers and the townsmen, after three days' negotiation, agreed to give up the town of New Amstel to the English; but D'Hinoyossa and the soldiers refused to surrender, and they retired into the Fort. The following were the terms of capitulation, viz:

[1] Register of Pennsylvania, vol. 1, p. 37; New York Records.

"1. That all the burgomasters and planters will submit themselves to His Majesty's authority, without making any resistance.

"2. That whoever, of what nation soever, doth submit to His Majesty's authority, shall be protected in their estates, real and personal whatsoever, by His Majesty's laws and justice.

"3. That the present magistrates shall be continued in the offices and jurisdiction, to exercise their civil powers as formerly.

"4. That if any Dutchman or other person shall desire to depart from this river, that it is lawful for him so to do, within six months after the date of these articles.

"5. That the magistrates and all the inhabitants (who are included in these articles) shall take the oath of allegiance to His Majesty, and of fidelity to the present governor.

"6. That all the people shall enjoy the liberty of their conscience in church discipline as formerly.

"7. That whoever shall take the oath is from that time a free denizen, and shall enjoy all the privileges of trading into any of His Majesty's dominions as freely as any Englishman, and may require a certificate for so doing.

"8. That the schout, the burgomaster, sheriff, and other inferior magistrates shall use and exercise their customary power in administration of justice within their precincts for six months, or until His Majesty's pleasure is further known.

"*The Oath.*—I do swear by the Almighty God that

I will bear faith and allegiance to His Majesty of Great Britain, and that I will obey all such commands as I shall receive from the governor, deputy governor, and other officers appointed by His Majesty's authority, so long as I live within these or any other of His Majesty's territories.

"Given under my hand and seal this first day of October, in the year of our Lord God, 1664.

"ROBERT CARR.

"Given under our hands and seals, in behalf of ourselves and the rest of the inhabitants, the first of October, in the year of our Lord God, 1664.

"FOB OUT HOUT,	HANS BLOCK,
HENRY JOHNSON,	LUCAS PETERSON,
GERRITT S. VAN TIEL,	HENRY CASTURIER."

D'Hinoyossa having refused Carr's proposition to surrender peaceably, and having retired with the soldiers into Fort New Amstel, and it is believed with Alricks and Van Sweringen with him, Carr proceeded to use forcible means. Accordingly upon the Sunday morning following he landed his troops, and commanded his ships to fall down below the fort, although within musket shot, and to fire into it two broadsides each. This was done. The ships fired into Fort New Amstel, and the land troops making an attack at the same time took it by storm. The Dutch lost three men killed and ten wounded in this affair. After the fort was taken, the soldiers and sailors commenced to plunder, and succeeded in getting a great deal of booty. The noise and confusion was so great during this interval, that "no words of command could be heard for some time." Carr did his utmost to prevent this, and keep as many of the goods as he could entire. Fort New Amstel, though mounting fourteen guns, "was not tenable."[1]

After the capture of the town and Fort of New Amstel, a general scene of plunder took place. All the soldiers and many of the citizens of New Amstel were sold as slaves to Virginia (for white slavery or forced service then existed, as well as black). The negroes brought by the Gideon, and run across New Jersey by Alricks (as well as more than probably others, that could be found) were forfeited, and mostly divided amongst the captors, save those that the Dutch managed to conceal. Several were taken belonging to Alricks. Eleven were returned to him some four years afterwards by Ensign[2] Arthur Stock as a free gift.[3] They also took from the Dutch all the produce of the land for that year, and amongst other things were 100 sheep, 30 or 40 horses, 50 to 60 cows and oxen, a brew-house and still belonging to it, and a saw-mill ready to put up. (This is the first mention we have of a saw-mill in Delaware). They also plundered the settlement of the Mennonists at the Hoernkill, leaving the inhabitants there (to use the words of Van Sweringen) "not even a

[1] See Carr's letter, London Documents, vol. i. p. 204.

[2] Ensign in the English service is the lowest commissioned officer in the company.

[3] In this gift there was some act of policy, the reason of which is not at this time perfectly known.

nail."[1] Stuyvesant also in writing of this affair says: "That although the citizens of New Amstel made no resistance, 'they were stripped' and 'utterly plundered.'" He also confirms the selling of the citizens and soldiers as slaves. The amount of plunder obtained amounted to £4000. Carr, notwithstanding the amount of sheep and cattle taken from the unfortunate citizens of New Amstel, in writing to Colonel Nichols giving an account of the expedition, says: "That nothing was to be had on the Delaware but what was purchased from other places, and that to supply the wants of the garrison he had to send into Maryland some negroes belonging to D'Hinoyossa, which he sold for 'beef, pork, and salt,'" and, to use his own words, "other small conveniences," which, he said, "the place affordeth not."

Carr also complained of the Seneca and Tuscarora Indians, whom he said, "were exasperated by some Dutch and their own inclinations," and who "did violence both to heathens and Christians;" for which "the Indians of the neighborhood were unjustly blamed." Several murders, he said, "had been committed by them upon the Dutch and Swedes in less than six weeks." They were so strong on the east side of the river, "that no one dared to plant there."[2]

[1] From this it appears the Mennonists did settle at the Hoernkill (Lewistown). For Van Sweringen's account, see Broadhead and O'Calligan, vol. 3, pp. 343–6. He there alleges that the Schuylkill derived its name (sculk, hidden; kill, creek) from the Swedish vessel, Mercurius, that run past the batteries hiding there. See pp. 274–5, ante.

[2] London Documents, vol. i. p. 204.

He wished a treaty of peace to be made with them. The Guinea was immediately afterwards ordered to proceed with despatches to England.

In all the previous conflicts between the Dutch and Swedes, and Dutch and English, no life was lost, and no blood was shed, that history records, save in the cracking of the crown of an unfortunate white named Ever Ducking, whom some Englishman knocked over the head in a dispute about some land, between them and the Dutch, on the Fresh (Connecticut) river. This unusual violence was duly noted and denounced in the Dutch chronicles. Fort Cassimer was taken twice in the conflicts between the Dutch and Swedes, without a scratch being suffered by any one. Fort Christina was captured, and notwithstanding the length of the siege nobody was hurt. It was the same with Fort Amsterdam and Manhattan, which was surrendered without the least damage being done to any individual. So that the northern suburb of the pretty town of New Castle was the first and only place that was soiled in these conflicts with the stain of human gore. The encroaching Delaware, however, has since washed it away, and of the fort which was the scene of the carnage, and of the ground on which it stood, not a vestige now remains: all has been swallowed up in its waters. The site of Fort Cassimer, or New Amstel, as it was afterwards called, must have been a point of land on the northern side of New Castle. It then extended probably over a quarter of a mile further into the Delaware than now, as the river has washed and still washes away from three to five feet annually. During every storm the skulls and bones interred in an old graveyard (which must have been in the rear of this fort) are now torn by the angry Delaware from the graves in which they were laid, and strewed along its shores.

On the 7th of November, 1676, or only twelve years after its capture by the English, the fort was in ruins. It was then granted to Englebert Lott. The following is the extract from the New Castle records authorizing its destruction:—

"7th of November, 1676, Englebert Lott preferring in Court a petition desiring a grant from this worpfl. Cort, to take up ye Lott att ye Easte End of this Towne, where the old Forte formerly stood. The Court granted the petition his said request, hee leveling the old walls and buildings upon the same, according to his honor's, the governor's regulations."[1] Englebert Lott must have pulled down the ruined walls, as the Labadists, Dankers and Sluyter, on their visit to New Castle in 1680 say that the fort was "demolished."

Thus fell the Dutch power on the Delaware, and the Anglo-Saxon—that race that is more than probably destined hereafter to make its language, laws, manners, customs, and institutions those of the world,[2]

[1] New Castle Records.

[2] The author has come to the conclusion, after maturely considering the matter, that it is the destiny of the Anglo-Saxon race (no matter how Utopian or improbable it may seem) finally to unite the world together under one form of government, and thus do away with wars between the nations of the earth. In accordance with this

became the possessors of New Netherlands, which was the single gap in their possessions, that prevented

belief, in 1868 he wrote an essay to the London Cobden Club, competing for the gold medal offered by the late Mr. Cobden for the "best essay on the best way of developing improved political and commercial relations between Great Britain and the United States of America." In this essay he recommended the re-uniting together of the two great branches of the Anglo-Saxon race (*i.e.* the United States and Great Britain and her colonies) into one government; thus forming a great Anglo-Saxon Confederacy on the following plan, viz: Great Britain to abandon the government of Ireland and her colonies, and leave them to manage their local affairs by legislatures chosen by themselves. The United States and Great Britain and her colonies then to be divided into representative districts of equal population; each district to elect a member to a legislature (to be composed either of one or two houses), to meet at a place chosen for a capital, to pass laws to govern the great united nation. All members, both of the general and state legislatures, to be elected by universal suffrage, and representation always to be in proportion to population. An executive or executives to be elected to govern the great nation, with similar powers to the President of the United States, and courts to be instituted to try causes of difference that might arise between the various States, according to laws made and provided beforehand. Perfect freedom of trade to be between the various States. Education to be compulsory and universal, so that no one should "grow up by its ignorance to jeopardize the general welfare." Libraries to be established in districts convenient to and open to all, where the laws of the Great United Nation, and books that contained such information as would tend to the welfare of the citizens, should be deposited by the Government. All nations to be admitted into the Union upon application, and allowed every right enjoyed by the rest of the States, and representation in its legislature in proportion to the number of their people, provided they agree to abide by its laws, and teach the English language in their schools, so as to have one language for common use throughout the Confederacy: thus adopting for the Anglo-Saxon or English speaking races, and finally for the world, a form of government similar to that of the United States, save that senators should be in proportion to population, and be elected by universal suffrage, instead of two from each State, chosen by state legislatures as at present. The author

their owning the territory on the Atlantic seaboard from the 31st to the 46th parallel of latitude, or from

thinks that the uniting together of two such powerful nations on the principle of exact justice—every member comprising it having rights equal with the rest—would make a community too powerful for any nation or combination of nations to compete with in warfare. That nation after nation would join it, as they would then save the great expense of their army and navy (for the same military and naval orce would do for all), and the devouring conscription for enforced service in their armies, to which most civilized nations are now subject, to protect themselves against other nations having similar military systems. That this and the freedom for trade existing amongst us—a trade that would not be interrupted by hostile imposts or wars, but which would be allowed to flow in its natural channels, each section producing what it could with most advantage to itself, and exchanging with the others its surplus—would induce country after country to join us, until finally possibly all the nations of the earth would be admitted to our Union. The world would then be one nation, with one language, the English. Wars would exist no more; universal peace would prevail. The words of the prophet would be fulfilled. Swords would (metaphorically) be turned into plows, and spears into pruning hooks. There would be peace throughout the world. Utopian as this scheme may seem, it is just such men as you and I, reader (in the United States and Great Britain), who have power to say whether this shall be done. We have simply (enough of us) to manifest by our votes at the polls that this is our desire, and the thing can be accomplished, as both in America and Britain the people control the government. It is simply a matter of will, and the two people can say whether they prefer to unite themselves as one together, and live in harmony, deciding their differences by courts, on whose benches shall sit such judges as Hale and Mansfield or Marshall and Story, or fall out and kill each other by wholesale, and destroy each other's property. If the German and the Italian, notwithstanding the sanguinary battles that have occurred between their States, forgetting all past quarrels, so yearn for a unity of their race as to be willing to wade through seas of blood to accomplish it, why should not the Anglo-Saxon be willing to do peaceably, with none so bold as to say to him "nay," what they could only succeed in accomplishing by a costly expenditure of life and money. The author has treated these views more at length in an essay to the

the St. Croix, which flows past the northern boundary of Maine, to the St. Mary's, which forms the southern boundary of Georgia. Never perhaps was the taking of so trifling a fortress as New Amstel by so insignificant an armament productive of such momentous results. The capture of this, the last hold of the Dutch, consolidating the English possessions, caused our admixture from Maine to Georgia into one people, the grant of Pennsylvania to Penn and New Jersey to Berkley and Carteret, and their settlement with the English speaking races. This consolidation of territory also enabled us to show a united front to Great Britain, when we threw off her yoke, declared our independence, and formed the mighty Republic of the United States of North America, which who shall say hereafter may not be the United States of the World. What would have been the result had a single Dutch armed ship aided D'Hinoyossa against the frigate Guinea? It might have repulsed the attack, and whilst New Amstel stood the Dutch would have endeavored to have recovered Manhattan. If New Netherlands, or even the Dutch settlements on the South river, had divided New England from Maryland and the states south of her, as a consequence there would have been no Pennsylvania, or possibly New Jersey or New York, but a foreign

European Permanent League of Peace at Paris, whose sitting was put an end to by the Prussian war. He also laid them formally in a series of resolutions before the Pennsylvania Peace Society on the 27th of November, 1869, and before the Peace Union, New York, May 27, 1870.

people, speaking a different language and having different customs, severing communication between the English colonies north and south of the Delaware, and thus prevented not only united action, but probably even communication between them. Could the British colonies, thus separated, without the aid of Delaware, Pennsylvania, and New York, ever have hoped to have thrown off the yoke of the mother country? What the course of events would have been no man can say. But the capture of New Amstel was the cause of the settlement of the Delaware by the English and the foundation of Pennsylvania, by the uniting of the territory from Maine to Georgia under English rule, which caused the union of the colonies as states, and as a consequence the formation of the great American Republic, which has had and will still have an immense effect on the destinies of the world. Without the capture of this little fortress this might not have been. Therefore, never perhaps in the history of the world was the capture of so insignificant a stronghold productive of such momentous events for the benefit of mankind.

After the capture of New Amstel, the English confiscated the property of D'Hinoyossa, Van Sweringen, and Alricks. Part of D'Hinoyossa's property consisted of 150 acres of marsh land near the fort, was granted to Captain John Carr; another part described as a certain island in the Delaware river, called Swarten Natton Island, bounded on the north by Christina Kill, and on the west by Little Creek, containing 300 acres, was granted to Thomas Wollaston,

James Crawford,[1] Herman Otto, and Gerard Otto. To John Carr was also granted all the estate, both real and personal, of Gerrit Van Sweringen, amongst which was a house and ground in New Amstel. Peter Alricks' estate was granted to William Tomm, amongst which property was an island in the Delaware, about seven miles below New Castle.[2] The grants to John Carr were made "for good services in storming and reducing the fort." To William Tomm, for services on the Delaware.[3] No mention is made of the reason of the grant to the others. Alricks, however, soon succeeded in being taken into favor by the English, for on the 21st of November, 1665, he received a special license to trade and traffic with the Indians in and about the Whorekill.[4] He was also allowed to pass from New York to Delaware, and from thence to Maryland and return, with a servant and six horses;[5] and a few years later he was appointed one of the counsellors.[6]

D'Hinoyossa, Van Sweringen, and many other citizens of New Amstel, after the surrender retired into Maryland. Several of them settled permanently there, and there many of their descendants yet re-

[1] This was the first ancestor of the numerous and influential family of Crawfords in this State, many of whom yet reside in this county. Theodore F. Crawford, Esq., of Wilmington, is one of his descendants.

[2] The Labadists speak of Peter Alricks owning an island opposite Burlington, N. J.

[3] Delaware Records.

[4] We now use the English mode of spelling this place.

[5] MSS. in Reg. Penn., vol. 4, p. 75.

[6] Smith New Jersey, p. 52.

main. D'Hinoyossa settled on Foster's Island, in the Chesapeake, where he lived several years: it is attached to Talbot county, Maryland. In 1671 he petitioned the Maryland Assembly that he, Margaretta his wife, and his children, Alexander, John, Peter, Maria, Johanna, Christina, and Barbara, might be naturalized. It is said that he applied to the English for an office: it is certain he did for his forfeited estate, which is proved by the following letter, the last known on record that he has written. It was written from the house of Captain Thomas Howell, of St. Mary's, to Colonel Nichols, who was then Governor of what was New Netherlands. It says: "Your honor's very agreeable answer to my letters came safely here, and I learn from it that your honor is sorry for my loss. If your honor would please to console me therein, it can be done by giving me the rest of my lost estate; and could I get it back, I am resolved to live and die under your honor's government, yea, on the same conditions that I had from the City of Amsterdam. Meanwhile, should your honor incline thereunto, the answer should be sent to me at Captain Thomas Howell's, in Maryland, where I shall remain two or three months. Should these not be accepted by your honor, I would hereby respectfully request you to send me a letter under your honor's hand to His Highness the Duke of York, in order that I may take occasion to apply in London to His Highness aforesaid on the subject."[1] Nicholson

[1] Neil's Terra Mariæ, p. 163. Also Broadhead and O'Callighan, vol. 3, p. 83.

paid no attention to this respectful petition, and the gallant old soldier eventually returned to Holland, and entered the Dutch army, where he served in the wars between Louis XIV., King of France, and the Republic of Holland. He was one of the garrison of a fortress that surrendered to the French, after which, it is said, he ended his days in Holland.[1] Whatever may be said of his discretion in defending an untenable fort, he evidently displayed great bravery. His whole course shows him to have been a man of great nerve and action. Van Sweringen also became naturalized as a citizen of Maryland. After leaving Delaware he resided in the town of St. Mary's, in that State.

The name of the town of New Amstel was changed to that of New Castle, which name it has ever since borne. Altona again received its old name of Christina, and the great river—a part of which forms a portion of our State—and the great bay into which it flows, lost its name of South river, and has ever since been known by its English name of Delaware.

Hugh Hyde and Thomas Morley, the captains of the frigate, for their services in capturing New Amstel, had granted to them a place to hold as a manor, named Grimstead, at the head of the Delaware. It was called by the Indians Chipussen. They were to stock and people it in six years; otherwise their grant would be voidable. They were to be lords of the manor, and had the right of holding "Courts Leet." This was an ancient English court, held by

[1] London Documents. O'Calligan, vol. 2, p. 554.

lords of the manor, which has now even in that country fallen into disuse. It had jurisdiction from common nuisances and other offences against the public peace and trade, down to eaves' dropping (listening under the eaves of houses) and other trifling offences. Its business is now done by Courts of Quarter and General Sessions.

CHAPTER XXVI.

Title of English and Dutch to New Netherlands—English had equal rights—Character of the prominent Dutch officials who took part in Delaware affairs—Character of Stuyvesant—Kind to his friends—Energetic—Great tyrant—Partial and unjust judge—His persecution of Just Teunissen—Of Sabout Claessen—Of the Eight Men whom he imprisons—Banishes Kuyter and Melyn—Confiscates their property—Also that of Van der Capelle—Banishes Van der Donk and Gouvert Lockermans—Arrests Augustine Herman—Arrests Van Dincklage, vice-governor and judge—Puts him in guard-house—Persecutes people with false suits—His will law—His government a fraud—Notary fears to buy property, for fear it would be confiscated—Insults Van Dyck the fiscal—His councillors foreignors or men of bad character—Dissolves a convention with threats—His haughty message—Is opposed—Charter procured for New Amsterdam—Retires to his farm after capture by English—His life and death—Complaint of deputies of New Amsterdam to States General—Their names—Their complaints—Company make laws to suit themselves—Plunder settlers' goods—Of tyranny of Stuyvesant's government, under which life and property are not safe—Company's poverty cause them to oppress—Second remonstrance of the deputies—Caustic review of the administration of Kieft and Stuyvesant—High duties injure trade—Goods mostly smuggled—Kieft and Stuyvesant claim sovereign power—Kieft's plan to build a church—Takes advantage of drunkenness to obtain subscriptions—Money raised for school and poor spent by Company—Excise laid on wine and beer—Kieft uses such lofty language that the people cannot understand him—Public money placed with the Company's—Free negro slaves against law—Character of fiscal Hooyhens—Censure of Stuyvesant—He uses public money for private purposes—Wastes it in unnecessary councillors—Company's measures suspected—Fort a ruin—Money raised to repair it spent by Stuyvesant—His abuse—He is an unjust judge—Partial and oppressive—Abuses those who differ with him—Abuses the Eight Men—Threatens to kill those who appeal to Holland—He sells arms to Indians—Carries on all sorts of business—Carries on trade which he forbids to others on pain of death—His propensity to confiscate everything damages trade of New Amsterdam—Ships afraid to come on that account—Collecting debts from citizens, refuses to deduct what the Company owes them—Enters room of Van der Donk—Seizes draft of complaint of deputies—Arrests him—Arrests and guards by soldiers all who differ with him—Excludes Van Dyck from the council—Character of Stuyvesant's council—Of Brian Nuton, who loves and fears Stuyvesant, and always says "yes"—Of Adrian Keyser, who holds his tongue—Of the captains of the ships as councillors, whom Stuyvesant calls "a pack of thieves"—Of Van Dyck, "whose head is a trouble to him"—They are the judges that rule New York and Delaware—Bad character of Tienhoven, the Secretary of State.

It has been the custom of many writers to condemn the English for their attack upon the Dutch, and the capture of New Netherlands. The Dutch have been considered in the right, and the English in the wrong. It is not our purpose in this history to enter into an elaborate discussion of the title of the two nations to the territory on this continent then called New Netherlands. It will be simply sufficient to state that the English discovered and explored that portion of the continent in which New Netherlands was situated before the Dutch; that they actually made formal claim, and granted portions of it to their citizens before the Dutch entered either the Delaware or the Hudson; that they continually, both on the Delaware and Hudson, denied the right of the Dutch to the land, gave them warning repeatedly that they were trespassers on

English territory, and made constant endeavors to settle, both on the Delaware and at New York. That on the former river their citizens, as will be seen by the former pages of this history, were driven away, their trading houses burnt, and their settlements destroyed. That at the first, when formal notice was given to the Dutch by the English ambassador, they made evasive replies, instead of honestly claiming their rights. That up to the time of the capture of New Netherlands the English were unceasing and incessant in their demands for the territory, and never for an instant gave up the claim to what they considered their own. The English, therefore, taking all the circumstances into consideration, appear to have had at least as good a right to the territory as the Dutch, and in our opinion a better. In the capture of the Dutch possessions they did not, therefore, commence an entirely causeless war, but took what they believed was their own lands, and of which they believed the Dutch not only held violent and wrongful possession, but in addition committed outrages on their citizens by driving them away, when they went there for the purposes of trade and settlement.

We close the first volume of our history by a description of the characters of Stuyvesant and Kieft, the last Dutch governors of New Netherlands, of whom Delaware was a part; and also a description of the character of the ruling Dutch officials, who, although residents of New Amsterdam, were mixed up with our affairs, and really ruled the territory now composing our State. And also in the succeeding chapter a description of such Dutch or Knickerbocker families that were famous, and who settled in New Netherlands before its conquest by the English, and whom we can trace by the records as having descendants resident in Delaware, and who, being amongst the patriarchs of our State, have their blood flowing through the veins of thousands of our citizens, mixed with that of every other civilized nation under the sun. They with the Swedes and Finns, the Huguenot French and Protestant refugees, and a body of leading Irish citizens, who emigrated to Delaware about the years 1737 and 1745, engrafted on the English that came with Sir Robert Carr and William Penn, may be considered as the progenitors of a large portion of the citizens of Delaware. They were the fathers of intermingled tribes, which exist not only in Delaware, but are spread over all the States of the Union, where they have always taken leading parts in the public and private business of whatever State they may happen to be residents. In a great proportion of Delaware it is only recently that the current of modern emigration has began to flow. Previous to 1845 there were barely a dozen families of foreign birth resident south of the Appoquinemink. In the city of Wilmington and the hundreds of Brandywine and Christiana, there has always been a steady flow of emigration and an infusion of new blood. But in the lower part of New Castle and Kent and Sussex the citizens were until recently mostly born on the soil, and their descent could be traced to, comparatively speaking, few families, principally Dutch, English, and Irish Presbyterians, and French refugee Protestants, intermingled with a few Swedes and Finns. Therefore, in giving a sketch of these leading Dutch families, we really give a history of the ancestors, from whom a large number of our citizens are descended. Of the others we shall speak in their chronological order.

Of the characters of the prominent men who resided, and ruled our State, on the Delaware, our readers can form a pretty correct judgment from the account of their actions in our previous pages. But of those resident at the Manhattans, a further description of them will be needed to form a correct idea of their characteristics. This we have been enabled to give from the disputes that took place between them, which taking the part of remonstrances and petitions, and having been reduced to writing and filed: their quarrels have thus illustrated their history.

Stuyvesant (which, reduced to English, means holy saint) was the supreme governor of Delaware (the others merely sub-governors), and one of the greatest tyrants. With some good, he had nearly all the bad qualities that would render him unfit for a ruler. He was bold and energetic, kind to his friends, and probably as moral as most men of his day. He was an elder in the church, and, as far as known, a faithful outward performer of his religious duties; but here the list of his good qualities may be said to end. If the statements made to the Government of Holland

and the West India Company are correct, and we have no good reason to doubt them, for they are in part proved, and were made by the best citizens of New Amsterdam, he was one of the worst rulers of his age. Another of the many instances that can be pointed at to show how little man is to be trusted with power over his fellows. He cheated both the Company and the public. Money going into his hands, wrenched from the suffering citizens of Manhattan, who had been precipitated into an unjust war with the Indians by Kieft, his predecessor, and whose homes were destroyed and whose fields were ravaged, he diverted to his own private purposes. He sought eagerly for confiscations, and using his powers as a judge, under the forms of law robbed all who were opposed to him within his power. He was abusive to all with whom he differed, both in his public and private intercourse. We will cite the following as a few cases of his injustice, viz.:—Whilst he sold the guns and powder of the Company to the Indians, and kept the proceeds, on a similar charge being made, without proof, against Joost Teunissen, a baker, he threatened him with torture, and when he applied for permission to travel through the country to buy wheat to carry on his business, refused him a license and threatened him with "a caning;" and so malignantly persecuted him and Sabout Claessen that they had to fly from the colony. The latter he deprived of his property. Every man who opposed him he endeavored to crush and ruin. The Eight Men elected as the counsellors of the citizens who made the complaint against the Indian war, to use the words of Van der Donk, "he caused to be separated, put in prison, locked up, or hunted and utterly terrified." In their complaint they alleged, "the Indians lived amongst them like lambs, injuring no one, but affording them every assistance, until they suffered outrages which originated in a foolish hankering after war." He banished Joachim Peterson Kuyter and Cornelius Melyn—one, as he alleged, for shaking his finger at him in the council of the Eight Men, but the real reason for the banishment of both was sending Gouvert Lockermans to Holland with the complaints of the Eight Men in relation to his evil government. They appealed from his judgment to Holland, and in revenge he confiscated their property, and also confiscated a vessel belonging to Van der Capelle, the Patroon of Staten Island, on a false charge, because he thought that Melyn had in some way an interest in it. Another instance of his oppression was in the case of Van der Donk, who had prepared a complaint of the Select Men to send to Holland. He entered his room, seized on the rough draft of it, and banished him from the colony. He trumped up a charge against (and prosecuted capitally) Gouvert Lockermans, one of the Nine Men,[1] and sentenced him to three years' banishment, and threatened to enforce that sentence unless he signed a certificate that he could say nothing of him "but what was honest and honorable." He arrested Augustine

[1] A body of men chosen by the citizens of New Amsterdam to assist in the government.

Herman for refusing to produce a paper that was drawn up for circulation amongst the Nine Men. He sent a file of soldiers into the court, and arrested Van Dincklage, the vice governor, whilst he was sitting on the bench as a judge, and confined him several days in the guard-house, and then acting beyond the power delegated to him, turned him out of his office. He instituted numerous false suits to ruin people—he himself acting as judge—and confiscated their property. The government was administered by himself and a few sycophants whom he controlled. "His will was the law." Ordinances were made and enforced of which the community received no notice. He imitated in his petty government a royal state, and had a guard of halbadiers around him. Those who attempted to appeal to Holland from his judgment were fined and imprisoned. So great was the terror of him, that a notary public sent from Holland wrote back, that he could get no one to assist him to prepare his papers, and that he dared not purchase property for fear false suit would be entered against him, and that it would be confiscated. He was insulting and brutal to those officers who differed with him, as will be seen in the sequel by the formal complaint of the Deputies of the people to the States-General, and in his treatment of Van Dyck the fiscal, who by law had the direction of all actions, both civil and criminal, and who was entitled to a seat in the court. When he endeavored to exercise this right, Stuyvesant told him to "*get out*," and said, "Whenever I want you, I will call for you." He also degraded him by ordering him to "keep the hogs out of the fort"—work before done by a negro—and whenever he spoke to him or contradicted him, to use his own words, "*got a growl just as if he would eat him up.*" He selected his councillors too often from the worst and most ignorant people, from foreigners to the exclusion of his own countrymen, and from men dependent upon his bounty, or who were devoted to his will. When in 1653 a convention assembled and demanded that no new laws should be enacted and no officers appointed but by the consent of the people, he ordered them to separate on pain of punishment, and said to them, "We derive our authority from God and the Company, and not from a few ignorant subjects." He was, however, manfully opposed by Gouvert Lockermans, Augustine Herman, and other sturdy burghers of New Amsterdam, notwithstanding his threats and confiscations. They finally succeeded in getting a charter for New Amsterdam, giving the citizens a voice in the government, and thus checking his power. After the capture of New Netherlands by the English, he retired to his farm.

Stuyvesant was born in Holland in 1602. He entered the army, served in the West Indies, and lost a leg in the attack on the Island of St. Martins. He was also at one time governor of Curocoa, one of the West India islands. He died in New York in 1682, and was buried in St. Mark's Church.

The most caustic review of his actions, and the men who administered the government under him, was made in 1564 by the Deputies of the citizens of

Manhattan in two formal complaints to the States-General. Amongst them were Augustine Herman and Gouvert Lockermans, the ancestors of several of our Delaware families. In these complaints, amongst other matters they spoke of unsuitable government, onerous imposts and duties, and long continued war. They alleged that the Company had never adhered to the privileges of New Netherlands, but always altered them to suit its own convenience; that "a man was not master of his own vessel, but that the Company's soldiers were put on board, goods by force discharged from their warehouse were roughly used and robbed by the Company's servants." Or as they quaintly express it, "They *bite sharpe* and *carry away*." They alleged that under Stuyvesant's government a man was not *sure of either life or property*. "If," said they, "he but say anything displeasing or otherwise offensive to the governors. This tyranny," they said, "consisted mostly in arrests, imprisonments, banishments, confiscations, harsh prosecutions, blows, scoldings, reckoning half faults for entire ones, &c., and offering every one as many insults as they can invent." The poverty of the Company, they said, caused them to have recourse to various bad finesses, such as extortions and confiscations. They also alleged that the "high duties and confiscations made *with partiality* ruined the trade; that the principal portion of it was done by smugglers." To use their own words, "The duty is high. Of inspection and seizure there is no lack, and thus lawful trade is turned aside, except some little which is carried on only *pro forma* in order to push smuggling under this cloak." After enumerating many grievances, they said, "The people have been driven away by harsh and unwarrantable proceedings, and that the Company had instructed Kieft 'to pick out faults where none existed, and to consider a partial a complete error.'" They also said that Kieft and Stuyvesant alleged that they were the same as the prince in Netherlands, and claimed sovereign power; and that Stuyvesant alleged, in addition, that the "*prince was above the law*." They complained of the following plan taken by Kieft to build a church, which exemplifies the manners of that age. Said they, "We lacked money, and where was this to be got? It happened about this time that Evergardus Bogardus, the clergyman, gave in marriage a daughter by his first wife. The director thought this a good time for his purpose, and set to work *after the fourth or fifth drink;* and he himself setting a liberal example let the wedding guests sign whatever they were disposed to give towards the church. Each then with a *light head* subscribed away at a handsome rate, one competing with the other, and although some *heartily repented* when their *senses came back*, they were obliged nevertheless to pay. Nothing could avail against it. The church then was located in the fort, in opposition to every one's opinion. The honor and ownership of that work," they say, "must be inferred from the inscription, which in our opinion is somewhat ambiguous, and reads thus: Anno 1642, William Kieft, Directeur General, *heeft de gemeente desen temple doen bouwen.* But laying aside that, the people paid for the church."

They also alleged that "the plate had long been passed around for a common school, which has been built *with words;* for," they said, "as yet the first stone is not laid." The money, however, had all been spent. The money collected for the poor was also spent by the Company. Excises were levied on wine and beer, and when remonstrance was made to Kieft, "instead of relief they received a sharp reprimand and a written answer, which, as was his custom," they said, "he had couched *in so lengthy and diffuse a style,* that poor, humble people, such as are here, must inevitably commit mistakes regarding it. Money," they complained, "contributed by the people for public purposes was absorbed amongst the Company's property, and the children of certain free negroes were held in slavery,"—to use their own words—"contrary to all public law, that any one born of a free Christian mother should not be a slave." In Kieft's fiscal, Van der Hooykens, they said, "no confidence could be placed, in consequence of his drinking, in which all his science consisted."

Their censure of Stuyvesant was even more severe than that of Kieft. They alleged that he wasted the public money in unnecessary counsellors; that money raised for public, he employed for private purposes; that the Company's grain measure was suspected, but, said they, "who dare say so?" They complained terribly of his "*ill and spiteful language,* even to those who were officially brought to speak with him. If he were not in a good humor," said they, "they were berated as rascals, beer drinkers, &c. The fort," they alleged, "laid like a mole hill in a ruin, whilst he had spent the money raised for its repair by the people for other purposes."

They alleged he differed from Kieft in being "more active and malignant in looking up causes for prosecution against his innocent opponents; and that in his court he would browbeat and dispute and harrass one of the two parties, not as beseemeth a judge, but as a zealous advocate; and that on business before his council he would say, 'Gentlemen, this is my opinion. If any of you have aught to object to it, let him express it.' If any one then on the instant offer objection, his honor burst forth incontinently *in a rage*, and makes such a *to-do* that it is dreadful; yea, he frequently abuses the councillors as this and that in foul language, better befitting the fish market than the council; and if all this be tolerated, he will not be satisfied until he have his way." They alleged that when Kieft, his predecessor, was accused before him, he acted as his advocate, and spoke of the "Eight Men" as boorish brutes, threatened Melyn, who appealed from his decision, with having him "hanged on the highest tree in New Netherlands," and also of threatening those who appealed from his judgment to Holland with death.

They accused him of selling powder and arms to the Indians; of preventing people by threats from letting it be known how they were treated; of carrying on all sorts of business; of being a brewer, a farmer, a part owner of ships, a merchant; of having various stores of his own; of trading in contraband articles, and of forbidding "trade to others on pain of death," and then carrying it on himself. They said his promptness in confiscating caused great discontent amongst the inhabitants. "Scarce a ship," said they, "comes near the place that he does not look upon as a prize. Everywhere there is such an evil report that not a ship dare venture from the Carribee Islands." Again they complained that although the people were impoverished by the war, yet he collected rigidly the debts of the Company, and would not allow debts owing by the Company to the citizens as an offset. Those who would not follow his wishes, they said he denominated as "rascals, liars, rebels, and usurers." They recited at length his entry into the room of Van der Donk, his seizure of the draft of the complaint of the Deputies, and of his arrest and trial before the Supreme Council on a charge of having committed *crimen lesæ majestatis.* Those who took part in public affairs, they alleged, if they acted contrary to his will and pleasure, were persecuted, imprisoned, and guarded by soldiers. Amongst other instances of his tyranny, they recited the exclusion of Hendrick Van Dyck from the council board for the space of twenty-nine months. For this he gave an excuse, that "he could not keep a secret, but divulged whatever was done there." He also frequently declared that he was a "villain, a scoundrel, and a thief." Such was the character given of Stuyvesant by the Deputies of New Amsterdam. In the same paper they give the following as the character of the principal men of his council, who were also the Supreme Council of Delaware.

Brian Nuton, an Englishman, who commanded the soldiers, they describe as ignorant of the Dutch law and language, who dreaded Stuyvesant, and honored him as a benefactor. To everything "proposed by Stuyvesant, he would say yes."

Adrian Keyser they described sarcastically as "a man who had not forgotten much law." His saying was, that he let God's water run over God's field. "He," they alleged, "can say nothing, and dare not say anything." The captains of the ships had a vote in the council when they were ashore, but they asserted "Stuyvesant kept them so dependent that they dared not speak. He," they said, "once called them before the minister 'a pack of thieves.'"

Of Hendrick Van Dyck, who was also fiscal and commissary, whom Stuyvesant called a "villain, a scoundrel, and a thief," they spoke of "as a man wholly intolerable alike in words and deeds. What shall we say," they exclaim, "of one whose head is a trouble to him, and whose screw is loose, especially when it is surrounded by a little sap in the wood, which is no rare occurrence, as he is master at home." These men were the "supreme bench of justices" that ruled our state, as well as New York.

The Deputies were, however, most severe on the Secretary of State of New Netherlands, Cornelius Van Tienhoven. The following is their description of his character. They say: "A great deal might be said of this man—more even than we are able to et forth. For brevity's sake, however, we shall select here and there a few traits. He is crafty, subtle, intelligent, sharp-witted—good gifts when properly applied. He is one of those who have been longest in this country; is thoroughly acquainted with every circumstance relating both to the Christians and the Indians. With the Indians even he has run about like an Indian, with little covering and a patch before him, through lust for the prostitutes to whom he has ever been excessively addicted, and with whom he has so much intercourse, that no punishment or menaces of the director can drive him from them. He is a great adept at dissimulation, and even when laughing intends to *bite*, and professes the warmest friendship when he *hates the deepest.* To every one who has business with him—and there is scarcely one but has—he gives a favorable reply, promises assistance, and assists scarcely anybody, or leads them continually off on some course or the other, except the minister's friends. In his words and acts he is loose, false, deceitful, and given to lying; prodigal of promises, and when it comes to performances, nobody is home. The origin of the war was attributed principally to him and some of his friends. The director was led astray by his false reports and lies. Now if the voice of the people, by this maxim, be the voice of God, of this man hardly any good can with truth be said, and no evil concealed. With the exception of the director and his party, the whole community cries out against him, as a villain, a murderer, and a traitor; and that he must quit the country, or there will not be any peace with the Indians."

CHAPTER XXVII.

The Dutch patriarchs—Sketch of families descended from them—Huguenot French—The first Bayard—John Paul Jaquett—Johannes de Hayes—The first Statts and Comegys—Herman sick—His second wife—His daughter Margaretta, first described Delaware young lady—His death—Gov. Bassett possesses his mansion—Its destruction by fire—Tradition of Herman and his horse—A description of his descendants—Alricks' descendants—Gouvert Lockermans—Sketch of his life—Sketch of his descendants—End of first volume.

We conclude this last chapter of the first volume of our history by a short account of some of the families descended from the Dutch patriarchs who were residents of Delaware—whose blood now flows through the veins of thousands of our citizens, both in our own and other states. Of some of these old families every link can be traced in the chain of their descent, from the first ancestor to the present existing offspring. Amongst these are the descendants of Augustine Herman and Gouvert Lockermans. In others the links are broken, and we only know them from the similarity and peculiarity of their names, both Christian and surname. Oftentimes a child of each succeeding generation has received the name of its father or grandfather, and so it has been handed down, until many of our citizens bear the same name as their first ancestor, who emigrated here more than two hundred years ago.

A large proportion of our public men have always

been of Dutch descent, either by the father's or mother's side. Even after the conquest of the State by the English, for many years most of the principal magistrates and other public officers were Dutchmen. Amongst the numerous families who are in whole or in part descended from the Dutch patriarchs, in many cases mixed with Huguenot French, are the Oldhams (on the mother's side), the Van Dykes, the Vandegrifts, the Bayards (on the mother's side), the Alricks, the Statts, the Vandevers, the Harmans, the Comegys, the Vangezels, the Jaquetts, the Van Zandts, the Vances, the Hyatts, the Cochrans, the Fountains, the Le Counts, the Blackstones, the Kings, the Andersons, and others. There were also families of Van Dykes, Petersons, and Andersons, who were Swedes. Amongst those who derive their descents from the Huguenots and refugee Protestant French are the Bayards, the Bellvilles, the Bouchells, the De Hayes, and others. The Delaware Bayards are descended from Nicholas Bayard, who fled from France to Holland, and married Anneke, a sister of Stuyvesant. They had three sons, Belthazar, Peter, and Nicholas. Peter left New York, and came to Delaware with the Labadists. In 1675 he received a grant of Bombay Hook Island. Four years afterwards he purchased the right of the Indian owner, Maeesitt, Sachem of Canswick, for one gun and some other matters. From this Bayard it is believed the Bayards of Delaware are descended. Bayard street, in New York, is named after this family. The Bayards, like many of the other patriarchal Dutch

Huguenot families, have well maintained their social and political standing. The family have been distinguished for great talents. Three succeeding generations of them have represented the state in the United States Senate, viz.: the celebrated James A. Bayard, who signed the treaty of Ghent, then his sons, Richard Bayard and James A. Bayard, who sat there at different times, and Thomas F. Bayard, the son of the second James A. Bayard, who at the time of this writing represents the State in that body.

John Paul Jaquett, the second Dutch governor of Delaware, was also a French Protestant,[1] who had fled from France to Holland to avoid religious persecution. Before his arrival in Delaware, however, he had resided in Brazil. The Jaquetts lived on their farm, holding it from John Paul Jaquett, the first ancestor, until the time of the celebrated Major Peter Jaquett, the last surviving officer of the Revolution belonging to Delaware. This land was granted to Jaquett soon after the capture of Delaware by the Dutch. It is now called Long Hook, and belongs to Theodore Rogers, Esq. It is situated at the end of the causeway on the road from Wilmington to New Castle, about a mile from the bridge at the foot of

[1] The statement of Jaquett being a French Protestant is made on the authority of Miss Elizabeth Montgomery, in her "Reminiscences of Wilmington," a work written and edited without notes, and published in 1868. It gave a minute and graphic account of Wilmington, and its vicinity, and its citizens, and during her recollection. She was the daughter of Captain Hugh Montgomery, who was killed in a naval action during the Revolution. She was born in 1778, and died in the Episcopal Hospital, Philadelphia, a few years ago.

Market street, in that city. In 1699 the Labadists (Dankers and Sluyter) crossed the Christiana near to this farm. They speak of it as follows: "We proceeded thence a small distance overland to a place where the fortress of Christina had stood, which had been constructed and possessed by the Swedes, but taken by the Dutch governor, Stuyvesant, and afterwards, I believed, demolished by the English. * * * We were then taken over the Christina Creek in a canoe, and landed at the spot where Stuyvesant threw up his battery to attack the fort, and compelled them to surrender. At this spot there are medlar trees, which bear good fruit, from which one Jaquett, who does not live far from there, makes good brandy or spirits, which we tasted and found even better than French brandy."[1] By this it will be seen that in 1679 Fort Christina was destroyed.

From Johannes de Hayes are descended the Janvier (New Castle) family on the female side. A picture of him is still preserved in this family, and was exhibited to an audience during a lecture at Drawyers Presbyterian Church, in New Castle county, on the 10th of May, 1842, by the Rev. George Foote. Foote said, "He was evidently, as his costume shows, either a knight or a military officer of

[1] The medlar is a fruit not now raised in Delaware, or anywhere else in the Union that we know of. It is about the size of a peach, and is not eatable until perfectly rotten. Allusion is made to it in Shakespeare, when he says, "Like a medlar, rotten ere you are ripe." This is one of our extinct fruits. How many more plants were there that used to grow amongst us that are now extinct? The fig used to be raised in our gardens. There are none grown now.

high rank."[1] In 1676 he purchased of Joseph Chew, a farm of 400 acres, described in the New Castle Records as being near the "Old Landing" on the Appoquinimink Creek, for 2,000 pounds of tobacco, Dutch weight. He was then a merchant. He was afterwards a magistrate at New Castle, both under the Duke of York and William Penn.

After the capture of our state by the English, besides D'Hinoyossa and Van Sweringen, a number of other citizens of Delaware moved to Maryland. The principal evidence we have of this is the settling of so many Dutch and Germans in the neighborhood of the Sassafras and Bohemia rivers, and near the town of St. Mary's. They were, no doubt, brought there by the influence of Augustine Herman. Amongst these families who again settled in Delaware we are inclined to believe are the Comegys, the Cochrans, the Blackstones, the Le Counts, the Kings and others, and possibly the Bouchells. Several of them were naturalized by Maryland law from 1666 to 1684, amongst them were Peter Bayard, Arnoldus de la Grange, William Blackenstein (Blackstone), Hans Hanson, Cornelius Comegys, Gerritt Van Sweringen, besides Jacobson, Errickson, Peterson, and Le Count, whose Christian names are not given. In 1666 Augustine Herman petitioned the Maryland legislature for the naturalization of himself and all his

[1] The Rev. George Foote, a talented and highly esteemed minister of the Presbyterian Church, died at Odessa, in 1868. This lecture contains matter of much historical interest. It was published in 1842 in a small book.

family, viz.: Ephraim, Georgius, Gasparus, his sons, and Anna Margaretta, Judith, and Francina, his daughters.[1]

The Statts, now so numerous amongst us, were here as early as 1648. The first of them made mention of was Abraham Statts, surgeon, and elder of the church of Renslaerswick, New York. He was in 1651 driven from the island of Aharonnumy, in the Schuylkill, by the Swedes, and had his home burnt by the Indians in New York.

The first Comegys came from Vienna, in Austria. He was undoubtedly the ancestor of the present Comegys family. The late lamented Cornelius P. Comegys, who was governor of the State from 1836 to 1840, was undoubtedly a descendant of his, as he bears the same Christian name. One of his descendants (Joseph P. Comegys, son of the ex-governor) represented the State in the United States Senate. The Labadist, Dankers and Sluyter, give the following account of their visit to him in 1679. He is undoubtedly the Cornelius Comegys we have before spoken of as having been naturalized in Maryland. He appears to have been a man of wealth, owning several plantations, and employing several servants. He lived in Maryland, near the Sassafras river. They say: "We arrived at Cornelius, the son of Comegys, and called out to him, and he brought a canoe, which relieved us, as it was close on to evening. We thanked the person who had brought us, and stepped into the canoe. Cornelius, who was an

[1] Neil's Terræ Mariæ.

active young man, was pleased to meet Hollanders, although he himself was born in this country. We found Mr. Comegys on the next plantation, who bade us welcome; and after we had drank some cider, accompanied us with one of his company to Mr. Hosier's, who was a good, generous-hearted man, better than any Englishman we had met in this country. He had formerly had much business with Mr. Moll, but their affairs in England running behindhand a little, they both came and settled down here, and therefore Mr. Moll and he had a great regard for each other. * * * Mr. Comegys was from Vienna, and had a Dutch woman for a wife, who had taught her children to speak the Dutch language: they therefore had a kind disposition towards Hollanders. After her death he married an English woman, and he had himself learned many of the English maxims, although it was against his feelings; for we were sensible that he dared not work for us with an open heart. He told us that he would rather live at the Cape of Good Hope than here. 'How is that,' said I, 'when there is such good land here?' 'True,' he replied; 'but if you knew the people here as well as I do, you would be able to understand why.'"

Augustine Herman will hereafter cease to take part in Delaware history, save in a grant of land to the Labadists. Of all his children only the issue of his son Gasparus are now alive. From him are descended the Oldhams and the Bouchells. James R. Oldham, of Christiana Bridge, is the only male descendant now residing in the State. He is seventh in descent from the Augustine Herman. This is one of the few families that can be traced by their descent without a break in the link. The line runs thus: Gasparus Herman left issue a son named Ephraim Augustine Herman, who left a daughter Catharine, who married Peter Bouchell, a descendant on one side from Hendrick Sluyter, one of the founders of the Labadists. A gentleman named Joseph Enser or Inser married Mary, their daughter. They had one son, who was killed whilst celebrating his twenty-first birthday. He had given an entertainment to some young men, and whilst running races for amusement with their horses, he was thrown and killed. Colonel Edward Oldham, of the Maryland line of the Revolution, grandfather of J. R. Oldham, married their daughter Mary. There are several on the female side, both in Delaware and Maryland, descended from Colonel Oldham and Mary Ensor. In 1679 the Labadists visited Augustine Herman. They found him sick, and his family broken up from a termagent wife, who had driven his children away. They say: "He showed us every kindness he could in his condition, as he was very miserable, both in soul and body. His plantation was going much to decay, as well as his body, from want of attention. There was not a Christian man, as they term it, to serve him—nobody but negroes. All this was increased by a miserable, doubly miserable wife; but so miserable that I will not relate it here. All his children have been compelled on her account to leave their father's house. He spoke to us of his land,

and said he would never sell or hire it to Englishmen, but would sell it to us cheap if we were inclined to buy." At a second visit they describe his wife as the most artful and despicable creature that can be found. They also called Herman "a godless person." We must, however, receive with great allowance the account of the Labadists, who were a sour sect.

Augustine Herman died a short time after this, and was buried on the manor. His death must have occurred about the last of December, 1669, as on the 14th of December, after they left him, whilst visiting his son Ephraim, they were informed that he was very sick and at the point of death, and that his daughter Margaret had gone there to attend upon him in that condition.

The Bayards, who afterwards came into that portion of the manor on which was situated the grave of Herman, took the tombstone for a door for their family vault. The inscription on it is as follows:—"Augustine Herman, Bohemian, the first founder and seater of Bohemia Manor, Anno 1669." In this vault lies buried the remains of Richard Bassett, a former governor of Delaware, a member of the convention that formed the Constitution of the United States, and the father-in-law of the first James A. Bayard.

The following tradition is related of Herman, of which, however, we found no allusion to in the records, notwithstanding a careful search. An account will be found of it both in Ledmun[1] and Foot.

[1] Ledmun's Rise of Methodism in America.

Ledmun says: "It is said that the Dutch had him a prisoner of war at one time, under sentence of death, in New York. A short time before he was to be executed, he feigned himself to be deranged in mind, and requested that his horse should be brought to him in the prison. The horse was brought, finely caparisoned. Herman mounted him, and seemed to be performing military exercises, when on the first opportunity he bolted through one of the large windows that was some fifteen feet above ground, leaped down, swam the North river, run his horse through New Jersey, and alighted on the bank of the Delaware opposite New Castle, and thus made his ecsape from death and the Dutch. This daring feat, tradition says, he had transferred to canvas—himself represented as standing by the side of his charger, from whose nostrils the blood was flowing. It is said that a copy of this painting still exists. He never suffered this horse to be used afterwards, and when he died had him buried, and honored his grave with a tombstone."

The author has seen the copy of this painting. It is in the possession of James R. Oldham, Esq. It is as represented by Ledmun.

The old mansion house of Herman was occupied by Governor Bassett, and soon after his death it was burned down. Ledmun says: "Many old valuable paintings were consumed with this house. One of its large halls was lined with them. Many of them had belonged to Augustine Herman, the founder of Bohemia Manor. His likeness and that of his lady perished; also the painting representing the flight from the Dutch in New York by means of his famous war charger. There are still people living who saw these paintings again and again before they were destroyed." Ledmun also says: "Herman was the great man of the region: he had his deer park; he rode in his coach, driven by liveried servants."

We will close our history of the Herman family by an account of Margaret, his daughter, who is the first Delaware young lady of whom history records a description. The Labadists met her just before she left her brother Ephraim's to attend the deathbed of her father. They said, "She showed us much kindness. She was a little volatile, but of sweet and good disposition." Again speaking of her, they said, "She possesses a good disposition, although a little wild, according to the nature of the country. She complained that she was like a wild and desolate vine trained up in a wild and desolate country; that she had always felt an inclination to know more of God quietly, and to serve him. She treated us with great affection, and received thankfully and acceptably what we said to her." The Cochrans, now so numerous and influential, it is alleged, are descended from Derick Kolchman (now changed to Cochran), who was one of those engaged in founding the Labadist colony.

The Alricks, one of whom (Lucas Alricks, Esq., of New Castle hundred) holds the land on which he lives from his first ancestor, have from the time of the first governor of that name been numerous and influential. Their blood flows in the veins of large numbers of the most respectable citizens of Delaware and other States; for, like most old Delaware families, their descendants are scattered over most of the States of the Union.[1]

Of all the Delaware Knickerbocker families none that we know of have so complete a chain of descent as the offspring of the celebrated Govert Loockermans, the sturdy leader of the citizens of New Amsterdam, and colleague of Augustine Herman. From him the Loockermans of Dover are descended. One of his descendants still occupies the family mansion at Dover, which was built, in 1742, by Nicholas Loockermans. We insert it, as it not only shows the number of generations, link by link, that has existed in the State since its first settlement, but also the fortunes of a prominent and representative Delaware family.

Govert Loockermans, the progenitor of the Loockermans, came to New Amsterdam with Vouter Van Twiller, the director general or governor of New Netherlands, in the caravel St. Martin or Hope, commanded by Juriaen Blanck, in the month of April, 1633, from Holland, in the service of the West India Company. At the time of his arrival, he was aged about seventeen years. He married Maria Jansen (a daughter of Roelf Jansen and his wife Annetje or Anneke Jans, who, after the death of her husband, married the Rev. Everardus Bogardus), and was by

[1] Levi Alricks, Esq., of Philadelphia, is making a thorough research into the genealogy of the Alricks family.

that marriage brother-in-law of Oloff Stevenson Van Courtlandt, whose son founded the Van Courtlandt manor, in the State of New York; also of Jacob Van Couwenhoven. He filled some of the highest civil and military offices in New Amsterdam. He was despatched with Jan Davitz in May, 1664, across the Green Mountains by Stuyvesant to arrange peace with the Mohawk Indians. At Warrington he concluded a treaty with them. About the same period he commanded a small armed vessel. He drove the English from a fort they had erected up the Hudson river; also at the head of an armed force he surprised and utterly extirpated a tribe of hostile Indians on Staten Island, who had greatly annoyed and injured the settlers in New Amsterdam. It is said that the memory of this indiscriminate slaughter of this tribe of Indians, although approved by the popular sentiment of his day, occasioned him much disquietude of conscience, after his retirement from active life, in his last hours. He was despatched at one period of his life, at the head of an armed force, to expel the Swedes and English, who had encroached on territory claimed by the Dutch on the Delaware river, near the present city of Philadelphia.

Govert Loockermans was also a successful merchant and politician. He headed the popular party of New Amsterdam, known as the "country party," and resisted the dictatorial assumption of Stuyvesant, the hard headed and wooden legged leader of the court or administration party, by wresting from him for the people the right of representation in the council called the Schnepens, of which he was a member in 1657 and 1661. This bridled the prerogative claimed by Stuyvesant, and made the govern-ernment of the Manhattans in a measure, republican. He was three times banished by Stuyvesant, and was as often recalled on account of his public services. The feud between Stuyvesant and him was subsequently terminated by the marriage of his granddaughter with the grandson of Stuyvesant. After a career of honored usefulness, Govert Loockermans died in 1670, reputed the richest individual in North America. He was worth 520,000 Dutch guilders, an immense sum when the period in which he lived is considered. His public influence and position after his decease devolved on his son-in-law, Jacob Leisler, who became by a civil revolution the first governor of the people of the colony of New Amsterdam.

Govert Loockermans left five children, Elsie, Cornelis, Jacob, Joannes, and Maritjie. Elsie Loockermans married Cornelis P. Van-der-Veen, by whom she had Cornelis, Timothy, and Margaret. She next married Jacob Leisler, who subsequently acted so prominent a part in the early colonial history of New York.

Maritjie Loockermans married Balthazar Bayard, step-son to Governor Stuyvesant, and of this marriage was born Anna Maria, who married Augustus Jay, grandfather of Governor Jay. 2. Arietta, who married Samuel Verplank. 3. Jacobus, who married Hellegonda De Kay. 4. Judith, who married Gerardus Stuyvesant, grandson of the last Dutch governor, Peter Stuyvesant.

Joannes or Jannetje Loockermans was the second wife of Surgeon Hans Kiersted, and her children were Areantje, Cornelis, Jacobus, and Maritjie.

Cornelis Loockermans died, it is believed, childless in early life.

Jacob Loockermans, second son of the above named Govert Loockermans and Maria, his wife, was born A. D. 1650, in the city of New Amsterdam. He was a regularly graduated medical doctor, and practised medicine; but, he became a planter in 1682. On the 29th of January, 1677, he married Helena Ketin. Being involved in the political troubles, which culminated in the overthrow of his brother-in-law, Jacob Leisler,—(who was deposed and brought to the scaffold, by the royal governor of William III., of England), about the year 1681 he emigrated to Easton, in the State of Maryland, where he became a planter. He was a man of wealth, and left behind a great deal of real estate, in the city of New York, undisposed of. He died, on the 17th of August, 1730.

He left a son, Nicholas Loockermans, who was born, on the 10th of November, 1697. He married Sally Emerson, daughter of Vincent Emerson, of the Grange near Dover, in 1721. He died March 6, 1769, aged over seventy-one years. His tombstone remains at the Grange, to this day.

Vincent Loockermans, only child of the above named Nicholas Loockermans, was born at the Grange before mentioned, in 1722. He married as his second wife Elizabeth Pryor, daughter of John Pryor, merchant of Dover, in February, 1774. By his first wife Susannah, he had one child, Vincent Loockermans the younger. By Elizabeth Pryor, he had two children, viz.: Elizabeth and Nicholas. Vincent Loockermans the elder sat in the Legislature. He was a prominent Whig in the Revolution. He died at his residence, in Dover, on the 26th of August, 1785, in his sixty-third year. He left a large landed estate in and around Dover.

Nicholas Loockermans, son of Vincent Loockermans and Elizabeth, his wife, was born November 27, 1783. He sat in the Legislature, and died March 20, 1850. He was never married.

Elizabeth Loockermans, the only daughter of Vincent Loockermans and Elizabeth, his wife, was born December 23, 1779. She married Thomas Bradford, LL.D., of the city of Philadelphia, counsellor-at-law, the 8th of May, 1805. She died in the city of Philadelphia April 12, 1842, leaving her surviving, her husband and five children, viz.: Vincent Loockermans, Elizabeth Loockermans, Benjamin Rush, William, and Thomas Budd. She was buried along with her brother in her husband's family vault in the burial ground of the Second Presbyterian Church of Philadelphia, which vault has since been transferred to Laurel Hill, Philadelphia.

Vincent Loockermans Bradford, eldest surviving child of Elizabeth Loockermans and her husband, Thomas Bradford, was born September 24, 1808. He adopted the legal profession, and was admitted to practice, in Philadelphia, in April, 1829. He remeved to the State of Michigan in 1835, and was elected, in 1837, to the Senate of that State. He resumed the practice of his profession in Philadelphia in 1843, and was elected President of the Philadelphia and Trenton Railroad Company in 1859, being subsequently re-elected until 1871, inclusive. He married July 21, 1831, Juliet S. Rey, daughter of Emanuel Rey, Esq., planter of the Island of St. Martin, West Indies. He still lives.

Elizabeth Loockermans Bradford, eldest daughter of Elizabeth Loockermans and her husband, Thomas Bradford, married the Rev. William T. Dwight, D.D., of Portland, Me. (a son of Timothy Dwight, D.D., the distinguished President of Yale College). She died in 1863. Her husband died in 1865. She left surviving four children, the Rev. Henry E. Dwight, M.D., Thomas Bradford Dwight, counsellor-at-law, of Philadelphia, Elizabeth Bradford Dwight, and Mary W. Dwight—all of whom are now alive.

Benjamin Rush Bradford, of New Brighton, Beaver county, Pennsylvania, son of Elizabeth Loockermans and her husband, Thomas Bradford, Jr., married in 1860 Margaret Campbell, of Butler county, Pennsylvania. They have four children, viz.: Juliet S., Thomas, Eleanor C., and William C., all of whom now live.

William Bradford, of the city of Philadelphia, son of Elizabeth Loockermans and Thomas Bradford, was born in 1815. He still lives.

Thomas Budd Bradford, son of Elizabeth and Thomas Bradford, Jr., was born in 1816. He is a minister of the gospel, and now resides in the ancestral mansion of the Loockermans at Dover, which has sheltered the blood for more than a century. He still farms as proprietor much of the old Loockermans land contiguous to Dover. By his first wife he had no issue. He married as his second wife Miss Lucy H. Porter, a daughter of Dr. Robert R. Porter, an esteemed and influential citizen of Wilmington, Delaware, a granddaughter of the Hon. Willard Hall, District Judge of the United States District Court of Delaware, and a great-granddaughter of Chancellor Killen, of Delaware. His issue by this last marriage is four sons and one daughter. Since the foregoing was penned, Rev. Thomas B. Bradford departed this life, at Dover, March 25th, 1871.

A granddaughter of Vincent Loockermans the elder, by his first marriage, (being a daughter of Vincent Loockermans the younger), Elizabeth Loockermans, married Thomas Davy, of Philadelphia. She and her husband are both dead, leaving an only child, Mary S. Davy. Another grandchild of Vincent Loockermans the elder, by his first marriage, (being a daughter of Vincent Loockermans the younger) married the Hon. Nicholas G. Williamson, for many years Postmaster and Mayor of Wilmington; by whom she had issue, Mary Ann (married to Rev. Corry Chambers), Harriett (married to Hon. William D. Baker), Sallie E. (married to the Hon. Horn R. Kneas), Evelina (married to Courtlandt Howell, Esq.), Helena, and Elba (married to Leonard Phleger, Esq.).

Although the family, for a century past, have signed themselves and been called "Lockerman," the true spelling, as derived from the early records of the family, is "Loockermans."

It will be seen by this history of the descendants of Govert Loockermans how the blood of the Knickerbocker patriarchs is mingled and scattered over all the States, how the families maintain their position, and that seven generations of the descendants of the Locokermans and eight of the Hermans (for some of the last named descendants of both families have living children) have existed since the first settlement of Delaware. And as the same rule exists in all the families, we may consider from seven to nine generations of people have dwelt and now dwell in our State since the first white man took up his habitation upon our shores. The first volume of the first history of Delaware is now finished, and we hope that a kind Providence will allow us also to complete the second.

END OF VOL. I.

No. 1.

A HISTORY

OF THE

State of Delaware,

FROM

ITS FIRST SETTLEMENT UNTIL THE PRESENT TIME,

CONTAINING

A FULL ACCOUNT OF THE FIRST DUTCH AND SWEDISH SETTLEMENTS,

WITH

A DESCRIPTION OF ITS GEOGRAPHY AND GEOLOGY.

BY

FRANCIS VINCENT,

WILMINGTON, DEL.

PHILADELPHIA:

JOHN CAMPBELL, No. 740 SANSOM STREET.

1870.

HENRY B. ASHMEAD, Book and Job Printer, Nos. 1102 & 1104 Sansom St., Philadelphia.

A HISTORY

OF THE

STATE OF DELAWARE.

PUBLISHED IN NUMBERS.

THIRTY-TWO PAGES EACH.

Price, ... *30 cts. per Number.*

TO BE HAD OF

JNO. CAMPBELL, No. 740 Sansom Street, Philadelphia.

J. B. PORTER & E. S. R. BUTLER,

Or of the Author, Wilmington, Del.

Numbers sent post-paid, on the receipt of the money by the Author.

ADDRESS

FRANCIS VINCENT,

Wilmington, Del.

No. 15.

A HISTORY

OF THE

State of Delaware,

FROM

ITS FIRST SETTLEMENT UNTIL THE PRESENT TIME,

CONTAINING

A FULL ACCOUNT OF THE FIRST DUTCH AND SWEDISH SETTLEMENTS,

WITH

A DESCRIPTION OF ITS GEOGRAPHY AND GEOLOGY.

BY

FRANCIS VINCENT.

WILMINGTON, DEL.

PHILADELPHIA:

JOHN CAMPBELL, No. 740 SANSOM STREET.

SABIN & SON, No. 84 NASSAU ST., NEW YORK.

Or, can be had of the Author, in Wilmington, Del., who will mail it to any direction, on the receipt of the money. Price, 30 cts. per Number.

1871.

HENRY B. ASHMEAD, Book and Job Printer, Nos. 1102 & 1104 Sansom St., Philadelphia.

www.ingramcontent.com/pod-product-compliance
Lightning Source LLC
LaVergne TN
LVHW020922110826
845150LV00004B/742

* 9 7 8 1 4 2 5 5 5 4 5 2 1 *